Systematic List of Fossil Decapod Crustacean Species

Published in this series:

CRM 001 - *Stephan G. Bullard*	Larvae of anomuran and brachyuran crabs of North Carolina
CRM 002 - *Spyros Sfenthourakis et al. (eds.)*	The biology of terrestrial isopods, V
CRM 003 - *Tomislav Karanovic*	Subterranean Copepoda from arid Western Australia
CRM 004 - *Katsushi Sakai*	Callianassoidea of the world (Decapoda, Thalassinidea)
CRM 005 - *Kim Larsen*	Deep-sea Tanaidacea from the Gulf of Mexico
CRM 006 - *Katsushi Sakai*	Upogebiidae of the world (Decapoda, Thalassinidea)
CRM 007 - *Ivana Karanovic*	Candoninae (Ostracoda) from the Pilbara region in Western Australia
CRM 008 - *Frank D. Ferrari & Hans-Uwe Dahms*	Post-embryonic development of the Copepoda
CRM 009 - *Tomislav Karanovic*	Marine interstitial Poecilostomatoida and Cyclopoida (Copepoda) of Australia
CRM 010 - *Carrie E. Schweitzer et al.*	Systematic list of fossil decapod crustacean species

Author's addresses:
Dr. C.E. Schweitzer, Department of Geology, Kent State University Stark Campus, 6000 Frank Ave. NW, North Canton, OH 44720, U.S.A.; e-mail: cschweit@kent.edu
Dr. R.M. Feldmann, Department of Geology, Kent State University, Kent, OH 44242, U.S.A.; e-mail: rfeldman@kent.edu
Dr. A. Garassino, Museo di Storia Naturale, Corso Venezia, 55, I-20121 Milano, Italy.
Dr. H. Karasawa, Mizunami Fossil Museum, Yamanouchi, Akeyo, Mizunami, Gifu 509-6132, Japan.
Dr. G. Schweigert, Staatliches Museum für Naturkunde, Stuttgart, Germany.

Manuscript first received 13 April 2009; final version accepted 16 May 2009.

Cover: *Glyphea foresti* Feldmann & de Saint Laurent, 2002, from the Late Cretaceous of Australia.

Systematic List of Fossil Decapod Crustacean Species

By
Carrie E. Schweitzer, Rodney M. Feldmann,
Alessandro Garassino, Hiroaki Karasawa
and Günter Schweigert

CRUSTACEANA MONOGRAPHS, 10

BRILL

LEIDEN • BOSTON
2010

This book is printed on acid-free paper.

Library of Congress Cataloging-in Publication data

The Library of Congress Cataloging-in-Publication Data is available from the Publisher.

ISBN: 978 90 04 17891 5

PRINTED IN THE NETHERLANDS

CONTENTS

INTRODUCTION

Decapod crustaceans, the shrimp, lobsters, squat lobsters, and crabs have a geologic record extending from the Middle Devonian (Schram et al., 1978) to the present. Although they are rarely the most abundant fossils in an assemblage, they are important elements in many marine, aquatic, and terrestrial ecosystems. Most are opportunistic organisms, whose feeding patterns include scavenging, grazing, and predation. In addition to terrestrial and aquatic settings, modern decapods occupy nearly all marine habitats ranging from hydrothermal vent communities on the deep sea floor to intertidal and supratidal environments. They may occupy infaunal or epifaunal benthic settings or be nectonic. Because of this wide range of habitat preferences, studies of biodiversity and trophic organization would be incomplete without considering this species group.

Decapods are extremely rare in Paleozoic rocks, they increased in number and variety rapidly in the Mesozoic, and they exhibited great diversity in rocks ranging from Eocene to the present day. Fossil decapods have fascinated natural historians and paleontologists since the late 18th Century, and throughout the 19th Century the number of recognized genera and species increased rapidly. Inevitably, the limitations of literature availability and international communication resulted in a proliferation of names, many of which were later judged to be synonymous. The first major attempt to sort out synonymous names and to catalogue the then-known species was that of M. F. Glaessner's "Crustacea Decapoda", published in 1929 as Part 41 of Fossilium Catalogus I: Animalia, edited by J. F. Pompeckj. This monumental work recognized about 1349 species of fossil decapods arrayed in about 317 genera and stands as the primary entrée into older paleontological literature even today. This work provided ages, geographic distributions, classification, and synonymies of all known fossil decapods.

Placement of fossil decapods in higher taxa has been a straightforward exercise for those forms, primarily Cenozoic in age, that exhibit very close morphological affinities with extant forms. However, many ancient decapods are morphologically quite distinctly dissimilar from known modern animals and, therefore, they pose special problems in classification. Modern classification of Decapoda is largely based upon characters of the anatomy that only rarely are preserved in the fossil record, such as the maxillipeds and the reproductive structures. Therefore, it is necessary to employ morphological features that are preserved on fossils and relate those attributes to comparable features on extant forms. These so-called proxy characters (Schweitzer, 2003) provide a bridge that permits enriching our understanding of decapod history by placement of fossil taxa within classification schemes developed solely for extant forms. One effort to do so has been widely accepted and is cited in both paleontological and neontological studies, M. F. Glaessner's "Decapoda", published in 1969 as Part R, Arthropoda 4, Volume 2 of the "Treatise on Invertebrate Paleontology", edited by R. C. Moore. This work included a classification of genera into suprageneric ranks, but it does not present a comprehensive listing of species. The two works by Glaessner (1929, 1969) have formed the major bases for research on fossil decapods over the past four decades.

In the past four decades, the volume of literature on fossil decapods has grown at a rapid pace as has the study of living forms. As a result, the number of fossil decapods has increased enormously since 1929 and the classification of extant decapods has changed dramatically from the classification proposed by Glaessner (1969). Our current work, as of the first quarter of 2009, lists 3270 species of decapods arrayed within 916 genera and 178 families. An analysis of the history of classification is not relevant here, but suffice it that one recent classification has emerged as a platform for modern studies (Martin & Davis, 2001). Their classification was developed as a consensus arrangement resulting from consultation and input from carcinologists all over the world, and it appears to continue to be widely used. From the paleontological point of view, the classification suffers from not having incorporated taxa that are now extinct. At present, therefore, paleontologists studying decapods must rely on a massive, diffuse literature on generic and species occurrences published since 1929 and an excellent classification scheme that does not consider fossil taxa.

Because the classification of Martin & Davis (2001) did not include extinct taxa and because a large volume of systematic literature has been published subsequently, the present work reflects a significant modification of their classification. The purposes of this work are to present a compilation of all species of fossil decapods known to us, to place them in a classification framework substantially modified from that of Martin & Davis (2001), and to provide a complete list of references to the naming of species and higher level taxa employed in the classification.

Placement of fossil taxa into the modern classification of extant decapods has been done in two ways, based upon studies published in the last few decades. Many of the placements, including those of the Prosopidae *sensu* Glaessner, 1969 (cf. Feldmann et al., 2006; Schweitzer et al., 2007; Schweitzer & Feldmann, 2008b, 2008 [imprint 2007], 2009a-d) and the Raninoidea De Haan, 1841 (cf. Feldmann, 1992; Feldmann & Schweitzer, 2007; Feldmann et al., 2007), have been made using extremely detailed, but admittedly subjective, use of proxy characters. These placements have been examined by many paleontologists and have been sustained. Other assignments of fossil taxa have resulted from phylogenetic studies employing cladistic analyses. Amati et al. (2004) analysed characters from 31 fossil and extant decapods that established relationships within the Astacidea Latreille, 1802; Patek et al. (2006) developed phylogenetic hypotheses for the Palinura Latreille, 1802; Karasawa et al. (2008) placed the Portunoidea Rafinesque, 1815, within a phylogenetic context; Karasawa & Schweitzer (2006) re-evaluated the Xanthoidea MacLeay, 1838 *sensu lato*; and Karasawa & Kato (2001, 2003) performed phylogenetic analyses of the Grapsidae MacLeay, 1838, and Goneplacidae MacLeay, 1838, respectively. Although selection of characters for these analyses also introduces an element of subjectivity, the procedure is as rigorous as any currently available.

The classification of fossil decapod species must, of necessity, be considered a work in progress as research on the group increases. However, it is hoped that providing a comprehensive list of species and references to their naming, placed within a modern classification, will provide additional incentive for workers to continue research on this important group of organisms.

TAXONOMIC AND LITERATURE CONVENTIONS EMPLOYED

Conventions employed within the text. — Throughout the text, family names are printed in **bold face**. When the family and/or genus within a family contains fossil and extant species, the name of the taxon is followed by a dagger (†). When the taxon is represented only by extinct forms, it is denoted by a double dagger (††). Absence of the dagger symbol reflects a taxon known only from extant forms. All recognized families are placed within the classification and, in the event that there are no fossil representatives, that is so indicated. If all species within a genus are extinct, we did not place a double dagger after each species. Families, genera, and species are placed in alphabetical order in their respective lists.

The three numbers in parentheses following a generic name denote the number of species within the genus. The first number indicates the number of extant species; the second number indicates the number of species known from extant and fossil occurrences; and the third number indicates species known only from fossil occurrences. The tallies of fossil species are derived from this work: those of extant species are taken from various sources, including Baba et al. (2008) for the Galatheoidea; Ng et al. (2008) for the Brachyura; Patek et al. (2006) for some Palinura and Nephropidae; and unpublished estimates made by K. A. Crandall, S. De Grave, D. L. Felder, and J. W. Martin in generic-level compilations prepared as part of the NSF-funded AToL grant, "Morphological and Molecular Phylogeny of the Decapod Crustaceans".

Taxonomic decisions. — Every attempt has been made to assign species-level taxa to the genus currently held to be valid. Because systematic placement is subjective, and because evaluation of these taxa is an on-going process, it is only possible to indicate that placements herein represent our understanding of best current practice. The more important consideration is that the list of genera and species be as complete as possible. Individual taxa can then readily be re-evaluated and reassigned, if necessary, based upon new information.

Only one nomenclatorial decision has been made herein. The genus *Eoplax* Müller & Collins, 1991, is the junior homonym of *Eoplax* Ashby & Cotton, 1936. For this reason, we suggest the replacement name, ***Muelleroplax*** new name, for *Eoplax* Müller & Collins, 1991, in recognition of the contributions of Pál Müller, Hungarian Geological Institute, Budapest, to our understanding of decapod paleontology in Europe.

Authorships within the classification. — With few exceptions, the authorship of taxa has been verified by examination of the original work. Where that was not possible, primary literature was employed. The intent was to assign authorship as accurately as possible. For example, taxa named within the volume by Lőrenthey & Beurlen (1929) are frequently cited as having been authored by both; however, Lőrenthey was the author of most of the genera and species and the authorship herein is cited as Lőrenthey *in* Lőrenthey & Beurlen.

Determining the correct publication and date of publication, also presented challenges in several cases. The overarching principle was to assign as accurate a date of publication of taxa as possible while, at the same time, permitting recognition of the appropriate title for bibliographic purposes. For example, many of the taxa described by J. D. Dana are often attributed to his massive two-volume work, "United States Exploring Expedition during the Years 1838, 1839, 1840, 1841, 1842, under the Command of Charles Wilkes, U.S.N., Vol. XIII, Crustacea". The taxa discussed therein, however, were first described in a series of shorter works published in various outlets, and we have taken these to represent the actual dates of publication of the new taxa. Other works were published over a long period of time and the actual date of the part of the publication in which new taxa are reported are cited as, for example, H. Milne Edwards, 1834 [in 1834-1840]. In other cases, conflicting indication of dates has arisen because of the date of the imprint on the volume and the actual date of issue. We have adopted the following notation to avoid confusion: Bachmayer, 1953 [imprint 1951]. The actual date of appearance of the work was 1953 and that must be taken as the date of authorship. The imprint date is simply the date imprinted on the publication but in these cases is not the date of publication. Also noted was the attribution of the date when a paper was read as opposed to the date of publication of the proceedings volume. That was treated as follows: Stimpson, 1859 [read 1858].

Citations in the references. — Every attempt has been made to identify the precise work in which each of the taxonomic names were first named by using the same notations in the systematic list as in the references; that is, using 1953a, 1953b, when several papers by the same author appeared in the same year. When several multiple-authored papers appeared in the same year, enough of the authors are cited in the list to permit identifying the precise paper in the references.

For ease of locating authors with names bearing prefixes, all are alphabetized according to the first letter of the prefix. As much as could be determined, the prefixes were capitalized or set in lower case depending on what we determined was the usage of the original author. Thus, De Haan, De Man, d'Orbigny are placed in alphabetical order within the references within the letter "D." Similarly, Van Straelen and Von der Marck are arrayed within the letter "V." Yet, in order to facilitate retrieval of authors in the References, especially when it is not perfectly clear whether or not an author's name has a prefix, cross-references are added, as, e.g., "Straelen, V. Van *see* Van Straelen, V."

It should be noted, though, that capitalization of the prefix is different in the various countries: at least in Germany and the Netherlands, the prefix has a lower case initial when used in combination with the initial(s) of the author's forename(s) but has an initial capital when the surname is used without initials: e.g., in "W. de Haan" the prefix is written as "de", whereas in "(De Haan, 1833)" the "D" is capitalized. In Belgium, France, and Italy (at least), customs are different: there the prefix has a capital or lower case irrespective of the use of the initials: e.g., "S. De Grave". Thus, also "M. de Saint Laurent" remains "de Saint Laurent, M.", etc.

ACKNOWLEDGEMENTS

This work is a contribution to the group of principal investigators on the National Science Foundation Collaborative Research grant, "AToL: Morphological and Molecular Phylogeny of the Decapod Crustaceans" including K. A. Crandall, D. L. Felder, R. M. Feldmann, N. L. Hanegan, J. W. Martin, and C. E. Schweitzer. The suprageneric classification, originally derived from Martin & Davis (2001) has been vetted and augmented by this group, as well as by S. De Grave, Oxford University, and their contributions are gratefully acknowledged. Our sincere thanks to these individuals for their considerable contribution. Placement of extinct taxa within the classification, originally limited to extant taxa, and modifications to the Martin & Davis classification is the responsibility of Schweitzer and Feldmann. The comments of two anonymous reviewers as well as of De Grave on the manuscript were extremely helpful. The errors and omissions are solely the responsibility of the authors. Contribution supported by the Office of Research and Sponsored Programs and the Department of Geology, Kent State University, and NSF EF 0531670 to R. M. Feldmann and C. E. Schweitzer.

A CLASSIFICATION OF THE FOSSIL DECAPOD CRUSTACEANS

Please note that the indication "(type)" identifies the type-species of the genus at issue.

Order DECAPODA Latreille, 1802
Suborder DENDROBRANCHIATA Bate, 1888
Superfamily Penaeoidea Rafinesque, 1815
Family **Aegeridae**†† Burkenroad, 1963
Acanthochirana†† Strand, 1928 (=*Acanthochirus* Oppel, 1862 *non* Peters, 1861) (0, 0, 5)
A. angulata Oppel, 1862
A. cenomanica Glaessner, 1946
A. cordata (Münster, 1839) (type) (=*A. longipes* Oppel, 1862)
A. krausei Förster, 1967a
A. smithwoodwardi (Van Straelen, 1940a)
Aeger†† Münster, 1839 (0, 0, 20)
A. brevirostris Van Straelen, 1923a
A. brodiei Woodward, 1888
A. elegans Münster, 1839
A. elongatus Garassino & Teruzzi, 1990
A. foersteri Garassino & Teruzzi, 1990
A. franconicus Förster, 1980
A. gracilis Förster & Crane, 1984
A. hidalguensis Feldmann et al., 2007
A. insignis Oppel, 1862
A. laevis (Blake, 1876)
A. lehmanni (Langenhan, 1910)
A. libanensis Roger, 1946
A. macropus Garassino & Teruzzi, 1990
A. marderi Woodward, 1866a
A. muensteri Garassino & Teruzzi, 1990
A. robustus Garassino & Teruzzi, 1990
A. rostrospinatus Garassino & Teruzzi, 1990
A. spinipes (Desmarest, 1822)

A. straeleni Glaessner, 1929a
A. tipularius (Schlotheim, 1822) (type) (=*A. armatus* Oppel, 1862)
Aristeidae† Wood-Mason, 1891
Archeosolenocera†† Carriol & Riou, 1991 (0, 0, 1)
A. straeleni Carriol & Riou, 1991
Benthesicymidae† Wood-Mason, 1891
Benthesicymus† Bate, 1881 (16, 0, 1)
B. libanensis†† (Brocchi, 1875)
Carpopenaeidae†† Garassino, 1994
Carpopenaeus†† Glaessner, 1946 (0, 0, 3)
C. callirostris Glaessner, 1946 (type) (=*Pseudastacus dubertreti* Roger, 1946)
C. peterbuergeri Schweigert & Garassino, 2005a
C. septemspinatus (Dames, 1886)
Penaeidae† Rafinesque, 1815
Albertoppelia†† Schweigert & Garassino, 2004 (0, 0, 1)
A. kuempeli Schweigert & Garassino, 2004 (type)
Ambilobeia†† Garassino & Pasini, 2002 (0, 0, 1)
A. karojoi Garassino & Pasini, 2002 (type)
Antrimpos†† Münster, 1839 (0, 0, 15)
A. atavus (Bill, 1914)
A. crassipes (Bronn, 1858)
A. intermedius Oppel, 1862 (=*A. dubius* Münster, 1839; *A. tridens* Münster, 1839)
A. juvavensis Van Straelen, 1940a
A. kiliani Van Straelen, 1923a
A. latipes (Oppel, 1862)
A. liasicus (Oppel, 1862)
A. meyeri (Oppel, 1862) (=*A. senidens* Münster, 1839; *Koelga quatridens* Münster, 1839; *K. dubia* Münster, 1839; *K. tridens* Münster, 1839)
A. mirigiolensis Etter, 1994
A. monodon Münster, 1839
A. noricus Pinna, 1974
A. ? *secretaniae* Carriol & Riou, 1991
A. sharpi (Woodward, 1869)
A. speciosus Münster, 1839 (type) (=*A. angustus* Münster, 1839; *A. bidens* Münster, 1839; *A. decembens*

Münster, 1839; *A. trifidus* Münster, 1839; *Koelga quindens* Münster, 1839; *K. gibba* Münster, 1839; *K. septidens* Münster, 1839; *K. laevirostris* Münster, 1839)

A. undenarius Schweigert, 2001

Bombur†† Münster, 1839 (0, 0, 2)

B. aonis Bronn, 1858

B. complicatus Münster, 1839 (type)

Bylgia†† Münster, 1839 (0, 0, 4)

B. haeberleini (Münster, 1839)

B. ruedeli Schweigert & Garassino, 2004

B. spinosa Münster, 1839 (type)

B. hexadon Münster, 1839

Carinacaris†† Garassino, 1994 (0, 0, 1)

C. teruzzii Garassino, 1994 (type)

Cretapenaeus†† Garassino, Pasini & Dutheil, 2006 (0, 0, 1)

C. berberus Garassino et al., 2006 (type)

Drobna†† Münster, 1839 (0, 0, 1)

D. deformis Münster, 1839 (type)

Dusa†† Münster, 1839 (= *Palaeodusa* Pinna, 1974) (0, 0, 5)

D. araneae Schweigert, 2002

D. bronni Münster, 1839

D. denticulata Münster, 1839

D. longipes (Pinna, 1974)

D. monocera Münster, 1839 (type)

Hakelocaris†† Garassino, 1994 (0, 0, 1)

H. vavassorii Garassino, 1994 (type)

Ifasya†† Garassino & Teruzzi, 1995b (0, 0, 2)

I. madagascariensis (Van Straelen, 1933a)

I. straeleni Garassino & Teruzzi, 1995b (type)

Koelga†† Münster, 1839 (0, 0, 2)

K. curvirostris Münster, 1839 (type)

K. muensteri Schweigert & Garassino, 2004

Libanocaris†† Garassino, 1994 (0, 0, 1)

L. rogeri Garassino, 1994 (type)

Longichela†† Garassino & Teruzzi, 1993 (0, 0, 1)

L. orobica Garassino & Teruzzi, 1993 (type)
Macropenaeus†† Garassino, 1994 (0, 0, 1)
M. incertus (Roger, 1946) (type)
Microchela†† Garassino, 1994 (0, 0, 1)
M. rostrata Garassino, 1994 (type)
Micropenaeus†† Bravi & Garassino, 1998a (0, 0, 1)
M. tenuirostris Bravi & Garassino, 1998a (type)
Penaeus† Fabricius, 1798 (=*Pseudocrangon* Schlüter, 1862) (3, 0, 18)
P. arambourgi†† Roger, 1946
P. barmerensis†† Tiwari, 1963
P. bolcensis†† Secretan, 1975
P. cornappensis†† Garassino, 1998
P. furensis†† Garassino & Harper, 2004
P. glaessneri†† Tiwari, 1963
P. hamleti†† Garassino & Harper, 2004
P. kapurdii†† Prasad, 1966
P. maddeni†† Feldmann, Chirino-Galvez, Mason, Anderson, Duncan, Ward & Salem, 1993
P. natator†† Glaessner, 1946
P. obtusus†† Secretan, 1975
P. smyrnacus†† Van Straelen, 1940a
P. sorbinii†† Beschin & Garassino, 1999
P. spectabilis†† (Von der Marck, 1863)
P. tenuicauda†† (Von der Marck, 1858)
P. vanzii†† Beschin & Garassino, 1999
P. vernassensis†† Garassino & Teruzzi, 1995a
P. wenasogensis†† Rathbun, 1926a
Pseudobombur†† Secretan, 1975 (0, 0, 1)
P. nummuliticus Secretan, 1975 (type)
Pseudodusa†† Schweigert & Garassino, 2004 (0, 0, 1)
P. frattigianii Schweigert & Garassino, 2004 (type)
Rauna†† Münster, 1839 (0, 0, 1)
R. angusta Münster, 1839 (type)
Rhodanicaris†† Van Straelen, 1924 [imprint 1925] (0, 0, 1)
R. depereti Van Straelen, 1924 [imprint 1925] (type)
Satyrocaris†† Garassino & Teruzzi, 1993 (0, 0, 1)

S. cristatus Garassino & Teruzzi, 1993 (type)
Sicyoniidae† Ortmann, 1898
Sicyonia† H. Milne Edwards, 1830 (52, 0, 1)
S. roemeri†† (Von der Marck, 1858) (type)
Solenoceridae Wood-Mason, 1891
No fossil representatives
Superfamily Sergestoidea Dana, 1852a
Family **Luciferidae** De Haan, 1849
No fossil representatives
Sergestidae† Dana, 1852a
Cretasergestes†† Garassino & Schweigert, 2006a (0, 0, 1)
C. sahelalmaensis Garassino & Schweigert, 2006a (type)
Paleomattea†† Maisey & De Carvalho, 1995 (0, 0, 1)
P. deliciosa Maisey & De Carvalho, 1995 (type)
Suborder PLEOCYEMATA Burkenroad, 1963
Infraorder STENOPODIDEA Claus, 1872
Family **Macromaxillocarididae** Alvarez et al., 2006
No fossil representatives
Spongicolidae† Schram, 1986
Jilinicaris†† Schram, Yanbin, Vonk & Taylor, 2000 (0, 0, 1)
J. chinensis Schram et al., 2000 (type)
Stenopodidae† Claus, 1872
Phoenice†† Garassino, 2001 (0, 0, 1)
P. pasinii Garassino, 2001 (type)
Infraorder CARIDEA Dana, 1852a
Superfamily Procaridoidea Chace & Manning, 1972
Family **Procarididae** Chace & Manning, 1972
No fossil representatives
Family uncertain
Udora†† Münster, 1839 (0, 0, 5)
U. bigoti Van Straelen, 1924 [imprint 1925]
U. brevispina Münster, 1839 (type) (= *U. rarispina* Münster, 1839)
U. gevreyi Van Straelen, 1923a
U. koschnyi Schweigert & Garassino, 2004
U. minuta Van Straelen, 1923a

Superfamily Galatheacaridoidea Vereshchaka, 1997
Family **Galatheacarididae** Vereshchaka, 1997
No fossil representatives
Superfamily Pasiphaeoidea Dana, 1852a
Family **Pasiphaeidae**† Dana, 1852a
Pasiphaea† Savigny, 1816 (61, 0, 1)
P. mortua†† Smirnov, 1929
Superfamily Oplophoroidea Dana, 1852a
Family **Oplophoridae**† Dana, 1852a
Notostomus† A. Milne-Edwards, 1881a (9, 0, 1)
N. cretaceus†† Roger, 1946
Odontochelion†† Garassino, 1994 (0, 0, 1)
O. cretaceum (Roger, 1946) (type)
Oplophorus† H. Milne Edwards, 1837 [in 1834-1840] (4, 0, 3)
O. marcki†† Schlüter *in* Schlüter & Von der Marck, 1868
O. rosselli†† Vía, 1971
O. syriacus†† Roger, 1946
Tonellocaris†† Garassino, 1998 (0, 0, 1)
T. brevirostrata Garassino, 1998 (type)
Superfamily Atyoidea De Haan, 1849
Family **Atyidae**† De Haan, 1849
Caridina† H. Milne Edwards, 1837 [in 1834-1840] (272, 0, 1)
C. nitida†† A. Milne-Edwards, 1879
Delclosia†† Rabadà, 1993 (0, 0, 2)
D. martinelli Rabadà, 1993 (type)
D. roselli (Vía, 1971)
Superfamily Bresilioidea Calman, 1896
Family **Agostocarididae** Hart & Manning, 1986
No fossil representatives
Alvinocarididae† Christoffersen, 1986
Harthofia†† Polz, 2007 (0, 0, 2)
H. bergeri Polz, 2007
H. blumenbergi Polz, 2007
Bresiliidae Calman, 1896
No fossil representatives
Pseudochelidae De Grave & Moosa, 2004

No fossil representatives
Superfamily Nematocarcinoidea Smith, 1884
Family **Eugonatonotidae** Chace, 1937
No fossil representatives
Nematocarcinidae S. I. Smith, 1884
No fossil representatives
Rhynchocinetidae Ortmann, 1890
No fossil representatives
Xiphocarididae Ortmann, 1895
No fossil representatives
Superfamily Psalidopodoidea Wood-Mason & Alcock, 1892
Family **Psalidopodidae** Wood-Mason & Alcock, 1892
No fossil representatives
Superfamily Stylodactyloidea Bate, 1888
Family **Stylodactylidae** Bate, 1888
No fossil representatives
Superfamily Campylonotoidea Sollaud, 1913
Family **Bathypalaemonellidae** de Saint Laurent, 1985
No fossil representatives
Campylonotidae Sollaud, 1913
No fossil representatives
Superfamily Palaemonoidea Rafinesque, 1815
Family **Anchistioididae** Borradaile, 1915
No fossil representatives
Desmocarididae Borradaile, 1915
No fossil representatives
Gnathophyllidae Dana, 1852a
No fossil representatives
Hymenoceridae Ortmann, 1890
No fossil representatives
Kakaducarididae Bruce, 1993
No fossil representatives
Palaemonidae† Rafinesque, 1815
Alburnia†† Bravi & Garassino, 1998b (0, 0, 1)
A. petinensis Bravi & Garassino, 1998b (type)
Bechleja†† Houša, 1957 [imprint 1956] (0, 0, 4)
B. bahiaensis Martins-Neto & Mezzalira, 1991b
B. inopinata Houša, 1957 [imprint 1956] (type)

B. robusta Martins-Neto & Mezzalira, 1991b
B. rostrata Feldmann et al., 1981
Beurlenia†† Martins-Neto & Mezzalira, 1991a (0, 0, 1)
B. araripensis Martins-Neto & Mezzalira, 1991a (type)
Micropsalis†† Von Meyer, 1859 (0, 0, 2)
M. bolcensis Schauroth, 1865
M. papyracea Von Meyer, 1859 (type)
Palaemon† Weber, 1795 (42, 0, 6)
P. antonellae†† Garassino & Bravi, 2003
P. bahiaensis†† Beurlen, 1950
P. exul†† Fritsch, 1872
P. fabricii†† Michelotti, 1861
P. mortuus†† Smirnov, 1929
P. vesolensis†† Bravi et al., 1999
Propalaemon†† Woodward, 1903 (0, 0, 3)
P. longispinata Martins-Neto & Mezzalira, 1991b
P. minor Woodward, 1903
P. osborniensis Woodward, 1903 (type)
Pseudocaridinella†† Martins-Neto & Mezzalira, 1991b (0, 0, 2)
P. roxoi (Beurlen, 1950)
P. tremembeensis (Beurlen, 1950)
Schmelingia†† Schweigert, 2002 (0, 0, 1)
S. wulfi Schweigert, 2002 (type)
Yongjicaris†† Garassino, Yanbin, Schram & Taylor, 2002 (0, 0, 1)
Y. zhejiangensis Garassino et al., 2002 (type)
Typhlocarididae Annandale & Kemp, 1913
No fossil representatives
Superfamily Alpheoidea Rafinesque, 1815
Family **Alpheidae** Rafinesque, 1815
No fossil representatives
Hippolytidae Bate, 1888
No fossil representatives
Ogyrididae Holthuis, 1955
No fossil representatives
Superfamily Processoidea Ortmann, 1890
Family **Processidae** Ortmann, 1890

No fossil representatives
Superfamily Pandaloidea Haworth, 1825
Family **Pandalidae**† Haworth, 1825
Pandalus† Leach, 1814 (21, 0, 1)
P. orientalis†† Niino, 1951
Thalassocarididae Bate, 1888
No fossil representatives
Superfamily Physetocaridoidea Chace, 1940
Family **Physetocarididae** Chace, 1940
No fossil representatives
Superfamily Crangonoidea Haworth, 1825
Family **Crangonidae**† Haworth, 1825
Crangon† Fabricius, 1798 (19, 0, 2)
C. ? magnevillei†† Eudes-Deslongchamps, 1835
C. miocenicus†† (Garassino & Teruzzi, 1996) (=*Bannikovia*†† Garassino & Teruzzi, 1996; *Longitergite* Garassino & Teruzzi, 1996)
Morscrangon†† Garassino & Jakobsen, 2005 (0, 0, 1)
M. acutus Garassino & Jakobsen, 2005 (type)
Glyphocrangonidae Smith, 1884
No fossil representatives
Udorellidae†† Van Straelen, 1924 [imprint 1925]
Udorella†† Oppel, 1862 (0, 0, 1)
U. agassizi Oppel, 1862 (type)
Superfamily Pleopteryxoidea Schweigert & Garassino, 2006
Family **Pleopteryxidae**†† Schweigert & Garassino, 2006
Pleopteryx†† Schweigert & Garassino, 2004 (0, 0, 1)
P. kuempeli Schweigert & Garassino, 2004 (type)
Caridean Superfamily & Family uncertain
Acanthinopus†† Pinna, 1974 (0, 0, 1)
A. gibbosus Pinna, 1974 (type)
Alcmonacaris†† Polz, 2009 [imprint 2008] (0, 0, 1)
A. winkleri Polz, 2009 [imprint 2008] (type)
Blaculla†† Münster, 1839 (0, 0, 3)
B. brevipes Münster, 1839
B. nikoides Münster, 1839 (type)
B. sieboldi Oppel, 1862
Buergerocaris†† Schweigert & Garassino, 2004 (0, 0, 1)

B. psittacoides Schweigert & Garassino, 2004 (type)
Hefriga†† Münster, 1839 (0, 0, 3)
H. frischmanni Oppel, 1862
H. proboscideawulfi Schweigert & Garassino, 2004
H. serrata Münster, 1839 (type)
Gampsurus†† Von der Marck, 1863 (0, 0, 1)
G. dubius Von der Marck, 1863 (type)
Leiothorax†† Pinna, 1974 (0, 0, 1)
L. triasicus Pinna, 1974 (type)
Parvocaris†† Bravi & Garassino, 1998a (0, 0, 1)
P. samnitica Bravi & Garassino, 1998a (type)
Pinnacaris†† Garassino & Teruzzi, 1993 (0, 0, 1)
P. dentata Garassino & Teruzzi, 1993 (type)
Caridea incertae sedis
"*Dusa*" *bronni* Oppel, 1862
Infraorder GLYPHEOIDEA† Winkler, 1882
Superfamily Glypheoidea Winkler, 1883
Family **Chimaerastacidae**†† Amati, Feldmann & Zonneveld, 2004
Chimaerastacus†† Amati, Feldmann & Zonneveld, 2004 (0, 0, 1)
C. pacifluvialis Amati et al., 2004 (type)
Glypheidae† Winkler, 1883
Cedrillosia†† Garassino, Artal & Pasini, 2009a (0, 0, 1)
C. jurassica Garassino, Artal & Pasini, 2009a (type)
Glyphea†† Von Meyer, 1835 (=*Paraglyphea* Beurlen, 1928a) (0, 0, 53)
G. alexandri Taylor, 1979
G. ambigua (Von Fritsch, 1870)
G. arborinsularis Etheridge, Jr., 1917
G. australensis Feldmann et al., 1993
G. bathonica de Ferry, 1865
G. bicarinata Van Straelen, 1924 [imprint 1925]
G. bohemica Fritsch & Kafka, 1887
G. calloviensis H. Woods, 1927
G. ? *carinata* Tribolet, 1874
G. (?) *carolinensis* Rathbun, 1935a
G. carteri Bell, 1863
G. christeyi Feldmann & Maxwell, 1999

G. ? *couloni* Tribolet, 1874
G. crassa Oppel, 1861
G. cretacea McCoy, 1854
G. damesi Garassino, 2001
G. eureka Damborenea & Manceñido, 1987
G. ferruginea Blake & Hudleston, 1877
G. foresti Feldmann & de Saint Laurent, 2002
G. garasbaaensis Garassino, De Angeli & Pasini, 2008
G. georgianus Taylor, 1979
G. gussmanni Schütze, 1907
G. hennigi Beurlen, 1933
G. jeletzkyi Feldmann & McPherson, 1980
G. jurensis Oppel, 1861 (= *G. dietleni* Schütze, 1907)
G. liasina Von Meyer, 1840a
G. lyrica Blake, 1876
G. ? *mazetieri* Hée, 1924
G. ? *meyeri* Roemer, 1836
G. ? *meyeri* Tribolet, 1875
G. micheleae Schweitzer & Feldmann, 2001a
G. munsteri (Voltz, 1835) (= *G. speciosa* Von Meyer, 1835)
G. ? *neocomiensis* Robineau-Desvoidy, 1849
G. oculata J. Woods, 1957
G. prestwichi H. Woods, 1929
G. pseudoscyllarus (Schlotheim, 1822)
G. regleyana (Desmarest, 1822) (type) (= *G. martini* Étallon, 1861; *G. perroni* Étallon, 1861; *G. vulgaris* Von Meyer, 1836; *Palinurus longebrachiatus* Eudes-Deslongchamps, 1842)
G. reticulata Feldmann & Gaździcki, 1997
G. ? *riazi* Van Straelen, 1924 [imprint 1925]
G. rigoi Garassino, 2000
G. robusta Feldmann & McPherson, 1980
G. rosenkrantzi Van Straelen, 1929
G. serratosai Dupuy-Revilla, 1956
G. solitaria Oppel, 1861
G. ? *spinulosa* Van Straelen, 1924 [imprint 1925]
G. stilwelli Feldmann, 1993

G. ? *tantalus* Wöhrmann *in* Wöhrmann & E. Koken, 1892
G. tonelloi Garassino, 1997
G. tricarinata Garassino, 1996
G. vectensis H. Woods, 1927
G. viohli Polz, 2000
G. willetti (Woodward, 1878)
G. yoshiakii Kato & Karasawa, 2006

Litogaster†† Von Meyer, 1847a (0, 0, 6)
L. ? *luxoviensis* Étallon, 1859
L. obtusa Von Meyer, 1847a (type) [=*Brachygaster serrata* Von Meyer, 1847a; *L. venusta* Von Meyer, 1847a]
L. ornata Von Meyer, 1851a
L. tietenbachensis Assmann, 1927
L. tuberculata Assmann, 1927
L. turnbullensis Schram, 1971

Paralitogaster†† Glaessner, 1969 (=*Aspidogaster* Assmann, 1927) (0, 0, 2)
P. durlachensis (Förster, 1967b)
P. limicola (König, 1920) (type)

Squamosoglyphea†† Beurlen, 1930a (0, 0, 4)
S. redenbacheri Schweigert & Garassino, 2005b
S. rogeri Schweigert & Garassino, 2005b
S. squamosa (Münster, 1839)
S. udressieri (Von Meyer, 1840a) (type) (=*G. scabrosa* Philippi *in* Woodward & Salter, 1865; *S. squamifer* Eudes-Deslongchamps, 1842)

Trachysoma†† Bell, 1858 (=*Glypheopsis* Beurlen, 1928a) (0, 0, 15)
T. alpina (Oppel, 1861a)
T. antipodum (Glaessner, 1960)
T. bronnii (Vosinsky, 1848) (=*Glyphea vosinskyi* Lahusen, 1894)
T. cretacea (McCoy, 1849)
T. etalloni (Oppel, 1861a)
T. granulosa (Beurlen, 1928a)
T. heeri (Oppel, 1861a)

T. lundgreni (Schlüter, 1874)
T. ornata (Quenstedt, 1858)
T. pustulosa (Von Meyer, 1840a)
T. rostrata (Phillips, 1829) (=*Glyphea bradfordiensis* Hée, 1924; *G. hispida* Carter, 1886; *G. stonesfieldensis* Van Straelen, 1923b)
T. sanctaecrucis Collins, 1969
T. scabra Bell, 1858 (type)
T. terquemi (Oppel, 1861a)
T. ? *tomesi* (Woodward, 1868a)

Mecochiridae†† Van Straelen, 1924 [imprint 1925]
Huhatanka†† Feldmann & West, 1978 (0, 0, 1)
H. kiowana (Scott, 1970) (type)
Jabaloya†† Garassino, Artal & Pasini, 2009a (0, 0, 1)
J. aragonensis Garassino, Artal & Pasini, 2009a (type)
Mecochirus†† Germar, 1827 (0, 0, 17) (=*Mecochiria* Secretan, 1968)
M. chilensis Förster & Von Hillebrandt, 1984
M. clypeatus Carter, 1898
M. dubius (Münster, 1839)
M. eckerti Frentzen, 1937a
M. foresti (Secretan, 1968)
M. germari Garassino, 1996
M. grandis Quenstedt, 1856
M. houdardi Van Straelen, 1936a
M. leionoton (Krause, 1891)
M. longimanatus (Schlotheim, 1822) (type) (=*M. bajeri* Germar, 1827; *M. brevimanus* Münster, 1839; *M. locusta* Germar, 1827)
M. marwicki Glaessner, 1960
M. minimus (Hée, 1924)
M. olifex Quenstedt, 1856
M. pearcei McCoy, 1849
M. peytoni Woodward, 1876a
M. rostratus Collins & Rasmussen, 1992
M. socialis Von Meyer, 1841
Meyeria†† McCoy, 1849 (0, 0, 10)
M. bolivari Van Straelen, 1927

M. crofti Ball, 1960

M. gracilis Glaessner, 1932a

M. harveyi Woodward, 1900a

M. magna McCoy, 1849

M. mexicana Rathbun, 1935a

M. ornata (Phillips, 1829) (type)

M. pueblaensis Feldmann & Vega *in* Feldmann et al., 1995

M. rapax (Harbourt, 1905)

M. schwartzi Kitchin, 1908

? *Praeatya*†† Woodward, 1868a (0, 0, 1)

P. scabrosa Woodward, 1868a (type)

Pseudoglyphea†† Oppel, 1861a (=*Scapheus* Woodward, 1863; *Heteroglyphea* Colosi, 1921; *Triasiglyphea* Van Straelen, 1936b) (0, 0, 17)

P. alpina (Förster, 1971)

P. amalthea Oppel, 1861a

P. ancylochelis (Woodward, 1863)

P. etalloni Oppel, 1862

P. eximia Oppel, 1861a

P. foersteri Feldmann et al., 2002

P. friulana Garassino & Rigo, 2008

P. gigantea Garassino & Teruzzi, 1993

P. grandis (Von Meyer, 1837) (type) (=*P. winwoodi* Woodward, 1868b)

P. hamifera Tate *in* Tate & Blake, 1876

P. jourdani (Dumortier, 1867)

P. mulleri (Van Straelen, 1936b)

P. numismalis (Oppel, 1854)

P. paronae (Colosi, 1921)

P. spinosa (Assmann, 1927)

P. straeleni (Théobald, 1953)

P. terquemi Oppel, 1861a

? *Selenisca*†† Von Meyer, 1847b (0, 0, 1)

S. gratiosa Von Meyer, 1847b (type)

Pemphicidae†† Van Straelen, 1928a

Pemphix†† Von Meyer, 1840a (0, 0, 3)

P. malladai Vía, 1971

P. silesiacus Assmann, 1927
P. sueurii (Desmarest, 1822) (type)
Pseudopemphix†† Wüst, 1903 (0, 0, 3)
P. albertii (Von Meyer, 1840a) (type)
P. fritschi Wüst, 1903
P. meyeri (Alberti, 1864)
Sinopemphix†† Fenglin, 1975 (0, 0, 1)
S. guizhouensis Fenglin, 1975 (type)
Platychelidae†† Glaessner, 1969
Glaessnericaris†† Garassino & Teruzzi, 1993 (0, 0, 2)
G. dubia (Pinna, 1974)
G. macrochela Garassino & Teruzzi, 1993 (type)
Platychela†† Glaessner, 1931 (0, 0, 1)
P. trauthi Glaessner, 1931 (type)
Platypleon†† Van Straelen, 1936b (0, 0, 1)
P. nevadensis Van Straelen, 1936b (type)
Superfamily Erymoidea Van Straelen, 1924 [imprint 1925]
Erymidae†† Van Straelen, 1924 [imprint 1925]
Clytiella†† Glaessner, 1931 (0, 0, 1)
C. spinifera Glaessner, 1931 (type)
Clytiopsis†† Bill, 1914 (0, 0, 3)
C. argentoratensis Bill, 1914 (type)
C. audax (Von Meyer, 1834)
C. thuringica Förster, 1967b
Enoploclytia†† McCoy, 1849 (0, 0, 19)
E. armata Secretan, 1964
E. brevimana McCoy, 1846 (type)
E. collignoni Secretan, 1964
E. edwardsi Sauvage, 1891
E. eocenica Garassino, Artal & Pasini, 2009d
E. glaessneri Van Straelen, 1936a
E. granulicauda Schlüter, 1862
E. imagei McCoy, 1849
E. leachii (Mantell, 1822) (type) [=*E. heterodon* Schlüter, 1862]
E. minor Woodward, 1900a
E. paucispina Schlüter, 1862
E. salviensis (Robineau-Desvoidy, 1849)

E. sculpta Rathbun, 1926a
E. seitzi Glaessner, 1932a
E. triglypta Stenzel, 1945
E. tumimanus Rathbun, 1935a
E. tenuidigitata Woods, 1957
E. wenoensis Rathbun, 1935a
E. wintoni Stenzel, 1945

Eryma†† Von Meyer, 1840b (=*Erymastacus* Beurlen, 1928a; *Klytia* Von Meyer, 1840a) (0, 0, 59)
E. affinis de Ferry, 1865
E. americanum Rathbun, 1923a
E. amalthei (Quenstedt, 1851) [=*E. laedonensis* (Étallon, 1861); *E. propinqua* (Oppel, 1861b)]
E. anisodactylum (Krause, 1891)
E. australe Secretan, 1964
E. babeaui Étallon, 1861
E. bedeltum (Quenstedt, 1857) [=*Astacus birdii* Bean, 1839; *Glyphea aalensis* Quenstedt, 1857; *E. bizeti* (Morière, 1888); *Bolina etalloni* de Ferry, 1862; *Bolina girodi* Étallon, 1857; *E. greppini* (Oppel, 1861b); *Eryma wuerttembergica* Oppel, 1861b; *Eryma aspera* Oppel, 1861b; *Eryma elegans* Oppel, 1861b; *E. falcifera* (Morière, 1888); *Eryma corbieri* Morière, 1888; *E. gaiffei* Méchin, 1901; *E. lemairei* Méchin, 1901; *E. authelini* Méchin, 1901; *E. oppeli* Beurlen, 1928a]
E. boloniensis Sauvage, 1891
E. bordenensis (Copeland, 1960)
E. burgundiacum Crônier & Courville, 2004
E. caraboeufi Morière, 1888
E. compressum (Eudes-Deslongchamps, 1842)
E. corallinum Van Straelen, 1924 [imprint 1925]
E. crassimanum Krause, 1891
E. cretaceum Roger, 1946
E. cumonti Van Straelen, 1921a
E. curvum Beurlen, 1928a
E. delphinensis Moret, 1946
E. deslongchampsi Van Straelen, 1924 [imprint 1925]

E. dutertrei Sauvage, 1891
E. foersteri Feldmann, 1979
E. flectum Rathbun, 1926a
E. fossatum Krause, 1891
E. georgei (Carter, 1886)
E. gracilimanum Lahusen, 1894
E. granuliferum Secretan, 1964
E. guisei Wright, 1881 [1882]
E. hoerstgenensis Bachmayer & Malzahn, 1983
E. jungostrix Feldmann & Titus, 2006
E. karitzkyi Lahusen, 1894
E. lafayi Lissajous, 1923
E. lerasi Étallon, 1861
E. ? *leucodermum* (Pusch, 1838)
E. loryi Van Straelen, 1923a
E. madagascariensis Secretan, 1964
E. major (Oppel, 1861b)
E. mandelslohi (Von Meyer, 1840a)
E. meyeri Garassino, 1996
E. modestiforme (Schlotheim, 1822) (type; *non E. ventrosa* Von Meyer, 1835 *vide* Förster, 1966) [=*Glyphea laevigata* Münster, 1839; *E. leptodactylus* (Germar, 1827); *E. pseudoventrosa* Beurlen, 1928a]
E. mosquensis Lahusen, 1894 (=*Glyphea bronnii* Vosinsky, 1848, part)
E. nicklesi Méchin, 1901
E. nippon Karasawa, Ohara & Kato, 2008
E. ornatum (Quenstedt, 1857)
E. orthodactylum Beurlen, 1928a
E. pulchellum Carter, 1886
E. punctatum Oppel, 1861b
E. quadriverrucatum Trautschold, 1866
E. quenstedti Beurlen, 1928a
E. radiatum Oppel, 1861b
E. ? *rinellincolum* De Gregorio, 1884
E. romani Oppel, 1861b
E. squalidum Étallon, 1861

E. stantoni Rathbun, 1935a
E. strambergensis Bachmayer, 1959
E. stricklandi (Phillips, 1871)
E. sulcatum Harbort, 1905
E. tithonium Van Straelen, 1936a
E. tuberculatum Van Straelen, 1936a
E. ventrosum (Von Meyer, 1835) [= *E. thirriai* (Étallon, 1859); *E. subventrosa* Étallon, 1861; *E. thurmanni* (Étallon, 1861); *E. insignis* (Oppel, 1862); *E. villersi* Morière, 1883; *E. maeandrina* Krause, 1891; *E. morieri* Hée, 1924]
E. walkerae Feldmann & Haggart, 2007
Galicia†† Garassino & Krobicki, 2002 (0, 0, 3)
G. marianae Garassino & Krobicki, 2002 (type)
G. veltheimi (Münster, 1839)
G. westphali (Schweigert, Dietl & Röper, 2000)
Lissocardia†† Von Meyer, 1851a (non *Lissocardia* Von Meyer, 1847 n. n.) (= *Piratella*†† Assmann, 1927) (0, 0, 3)
L. badensis (Assmann, 1927)
L. magna Von Meyer, 1851a
L. silesiaca Von Meyer, 1851a (type)
Palaeastacus†† Bell, 1850 (0, 0, 23)
P. argoviensis Förster & Rieber, 1982
P. cenomaniensis Hée, 1924
P. decoratus (Frentzen, 1937b)
P. deslongchampsi Van Straelen, 1924 [imprint 1925]
P. dixoni Bell, 1850
P. edwardsi Étallon, 1861
P. falsani (Dumortier, 1867)
P. foersteri Taylor, 1979
P. fuciformis (Schlotheim, 1822)
P. kimzeyi Rathbun, 1935a
P. macrodactylus Bell, 1850
P. ? *ornatus* Whiteaves, 1887
P. ? *plavensis* Geinitz, 1875
P. poeschli Schweigert & Röper, 2001
P. rothgaengerae Schweigert & Röper, 2001

P. scaber (Bell, 1863)
P. selmaensis Rathbun, 1935a
P. spinosus (Étallon, 1861)
P. sussexiensis (Mantell, 1833) (type)
P. terraereginae (Etheridge, Jr., 1914)
P. triglyptus (Stenzel, 1945)
P. trisulcatus Schweitzer & Feldmann, 2001a
P. walkeri (Whitfield *in* White, 1883)
Paraclytiopsis†† Oravec, 1962 (0, 0, 1)
P. hungaricus Oravec, 1962 (type)
Protoclytiopsis†† Birshtein, 1958 (0, 0, 1)
P. antiqua Birshtein, 1958 (type)
Pustulina†† Quenstedt, 1857 (= *Phlyctisoma*†† Bell, 1863) (0, 0, 12)
P. calloviensis (Förster, 1966)
P. dawsoni (Woodward, 1900a)
P. elegans (Méchin, 1901)
P. granulata (Bell, 1863)
P. minuta (Schlotheim, 1822)
P. perroni (Étallon, 1861) [= *Enoploclytia dorsetensis* Woods, 1931]
P. pseudobabeaui (Dollfus, 1863)
P. scabra (Bell, 1863)
P. sinemuriana (Garassino, 1996)
P. spinulata (Secretan, 1964)
P. suevica Quenstedt, 1857 (type) (= *Palaeastacus solitarius* Oppel, 1862; *Eryma fraasi* Oppel, 1862)
P. tuberculata (Bell, 1863)
Stenodactylina†† Beurlen, 1928a (0, 0, 1)
S. liasina Beurlen, 1928a (type)
Incertae sedis
Eryma verrucosa (Münster, 1839)
Infraorder ASTACIDEA Latreille, 1802
Superfamily Palaeopalaemonoidea Brooks, 1962
Family **Palaeopalaemonidae**†† Brooks, 1962
Palaeopalaemon†† Whitfield, 1880 (0, 0, 1)
P. newberryi Whitfield, 1880 (type)
Superfamily Enoplometopoidea de Saint Laurent, 1988

Family **Enoplometopidae** de Saint Laurent, 1988

No fossil representatives

Uncinidae†† Beurlen, 1930a

Discussion. — Although considered a discrete Infraorder by Glaessner (1969), Uncinidae is herein placed within the Enoplometopidae following the analysis of Ahyong (2006). That analysis demonstrated that Uncinidae was the sister group to the Enoplometopidae.

Malmuncina†† Schweigert & Garassino, 2003 (0, 0, 1)

M. wulfi Schweigert & Garassino, 2003 (type)

Uncina†† Quenstedt, 1851 (0, 0, 4)

U. alpina Schweigert et al., 2003

U. ollerenshawi (Feldmann & Copeland, 1988)

U. pacifica Schweigert et al., 2003

U. posidoniae Quenstedt, 1851 (type)

Superfamily Nephropoidea Dana, 1852a

Family **Chilenophoberidae**†† Tshudy & Babcock, 1997

Chilenophoberus†† Chong & Förster, 1976 (0, 0, 1)

C. atacamensis Chong & Förster, 1976 (type)

Palaeophoberus†† Glaessner, 1932b (0, 0, 2)

P. portlandicus Roger & de Lapparent, 1945

P. suevicus (Quenstedt, 1867) (type)

Pseudastacus†† Oppel, 1861b (0, 0, 5)

P. hakelensis Fraas, 1878

P. minor Fraas, 1878

P. mucronatus (Phillips, 1836)

P. pusillus Van Straelen, 1924 [imprint 1925]

P. pustulosus (Münster, 1839) (type) (=*P. muensteri* Oppel, 1862)

Tillocheles†† Woods, 1957 (0, 0, 2)

T. kaoriae Yokoi & Karasawa, 2000

T. shannonae Woods, 1957 (type)

Protastacidae†† Albrecht, 1983

Protastacus†† Albrecht, 1983 (0, 0, 2)

P. antiquus (Harbort, 1905)

P. politus (Schlüter *in* Schlüter & Von der Marck, 1868)

Stenochiridae†† Beurlen, 1928a

Stenochirus†† Oppel, 1861a (0, 0, 3)

S. angustus (Münster, 1839)

S. mayeri Oppel, 1862 (type)
S. vahldieki Schweigert, Garassino & Riou, 2006
Nephropidae† Dana, 1852a
Cardirhynchus†† Schlüter, 1862 (0, 0, 1)
C. spinosus Schlüter, 1862 (type)
Homarus† Weber, 1795 (0, 2, 8) (number of extant species from Cobb & Castro, 2006)
H. americanus† H. Milne Edwards, 1837 [in 1834-1840]
H. brittonestris†† Stenzel, 1945
H. davisi†† Stenzel, 1945
H. hakelensis†† (Fraas, 1878)
H. lehmanni†† Haas, 1889
H. mickelsoni†† (Bishop, 1985a)
H. morrisi†† Quayle, 1987
H. neptunianus†† Polkowsky, 2004
H. travisensis†† Stenzel, 1945
H. vulgaris†† H. Milne Edwards, 1837 [in 1834-1840]
Hoploparia†† McCoy, 1849 (0, 0, 67)
H. albertaensis Tshudy et al., 2005
H. alpinus (Van Straelen, 1936c)
H. antarctica Wilckens, 1907
H. arbei Aguirre-Urreta, 1989
H. aspera Harbort, 1905
H. bearpawensis Feldmann, Bishop & Kammer, 1977
H. belli McCoy, 1849
H. benedeni Pelseneer, 1886a
H. bennetti Woodward, 1900a
H. beyrichi Schlüter, 1862
H. biserialis Fritsch & Kafka, 1887
H. blossomana Rathbun, 1935a
H. buntingi (Feldmann & Holland, 1971)
H. catalunica Garassino, Artal & Pasini, 2009b
H. calcarifera Schlüter, 1879
H. collignoni (Van Straelen, 1949)
H. columbiana Beurlen, 1938
H. corneti Van Straelen, 1921b
H. dentata (Roemer, 1841) (=*H. prismatica* McCoy, 1849; *Palaeno roemeri* Robineau-Desvoidy, 1849)

H. dentonensis Rathbun, 1935a

H. edwardsi (Robineau-Desvoidy, 1849) (=*H. blainvillii* Robineau-Desvoidy, 1849; *H. boscii* Robineau-Desvoidy, 1849; *H. cottaldi* Robineau-Desvoidy, 1849; *H. cuvieri* Robineau-Desvoidy, 1849; *H. desmarestii* Robineau-Desvoidy, 1849; *H. dorbignyi* Robineau-Desvoidy, 1849; *H. fabricii* Robineau-Desvoidy, 1849; *H. guerini* Robineau-Desvoidy, 1849; *H. herbsti* Robineau-Desvoidy, 1849; *H. lamarcki* Robineau-Desvoidy, 1849; *H. latreillii* Robineau-Desvoidy, 1849; *H. linnei* Robineau-Desvoidy, 1849; *H. lucasi* Robineau-Desvoidy, 1849; *H. michelinii* Robineau-Desvoidy, 1849; *H. neocomiensis* Tribolet, 1875; *H. sowerbyi* Robineau-Desvoidy, 1849; *Nephrops geoffroi* [*pars*] Robineau-Desvoidy, 1849; *Nephrops samviensis* [*pars*] Robineau-Desvoidy, 1849)

H. eocenica Lőrenthey *in* Lőrenthey & Beurlen, 1929

H. falcifer Fritsch & Kafka, 1887

H. fraasi (Böhm, 1891)

H. gabbi Pilsbry, 1901

H. gadzicki Feldmann & Crame, 1998

H. gammaroides McCoy, 1849 (=*H. belli* McCoy, 1849; *H. victoriae* Quayle, 1987)

H. georgeana Rathbun, 1935a

H. gladiator Pilsbry, 1901

H. groenlandica Ravn, 1903

H. hakelensis (Fraas, 1878)

H. hemprichi (Mertin, 1941)

H. heterodon (Bosquet, 1854)

H. horrida Schweitzer, Feldmann et al., 2003

H. intermedia Secretan, 1964

H. johnsoni Rathbun, 1935a

H. kamimurai Kato & Karasawa, 2006

H. kamuy Karasawa & Hayakawa, 2000

H. klebsi Noetling, 1885a (=*H. knetschii* Zimmermann, 1944 [imprint 1942])

H. lehmanni Haas, 1889

H. longimana (Sowerby, 1826) (type) (= *H. granulosa* Bell, 1863; *H. punctulata* Bell, 1863; *H. sulcirostris* Bell, 1863)
H. mcnairyensis Rathbun, 1926a
H. mesembria Etheridge, Jr., 1917
H. minima Tribolet, 1876
H. miyamotoi Karasawa, 1998a
H. muncki Pelseneer, 1886b
H. natsumiae Karasawa, Ohara & Kato, 2008
H. nephropiformis Schlüter, 1879
H. pelseneeri (Van Straelen, 1936a)
H. percyi Beneden, 1872
H. pusilla Secretan, 1964
H. riddlensis Feldmann, 1974
H. saxbyi McCoy, 1849
H. scabra Bell, 1863
H. schluteri Tribolet, 1876b
H. sculpta Secretan, 1964
H. senonensis Forir, 1887
H. seucica Schlüter, 1879
H. shastensis (Rathbun, 1929a)
H. stokesi (Weller, 1903)
H. sulcicauda Schlüter, 1879
H. tennesseensis Rathbun, 1926a
H. triboleti Borrisiak, 1904
H. trigeri (A. Milne-Edwards, 1886)
H. tshudyi Schweitzer & Feldmann, 2001a
H. victoriae Quayle, 1987
H. wardi Quayle, 1987
Jagtia†† Tshudy & Sorhannus, 2000 (0, 0, 1)
J. kunradensis Tshudy & Sorhannus, 2000 (type)
Metanephrops† Jenkins, 1972 (17, 0, 4)
M. elongatus†† Hu & Tao, 2000
M. jenkinsi†† Feldmann, 1989
M. motunauensis†† Jenkins, 1972
M. rossensis†† Feldmann et al., 1993
Nephrops† Leach, 1814 (1, 0, 7) (number of extant species from Bell et al., 2006)

N. aequus†† Rathbun, 1920a
N. americanus†† Rathbun, 1935a
N. costatus†† Rathbun, 1918
N. geoffroyi†† Robineau-Desvoidy, 1849
N. kvistgaardae†† Fraaije, Hansen & Hansen, 2005
N. maoensis†† Rathbun, 1920a
N. reedi†† Carter, 1898 (chela only, indeterminate)
Nephropsis† Wood-Mason, 1873 (16, 1, 1)
N. midwayensis†† Rathbun, 1935a
N. stewarti† Wood-Mason, 1873 (type)
Palaeonephrops†† Mertin, 1941 (0, 0, 2)
P. browni (Whitfield, 1907) (type)
P. westoni (Woodward, 1900a)
Paraclythia†† Fritsch & Kafka, 1887 (0, 0, 4)
P. boetticheri Mertin, 1941
P. nephropica Fritsch & Kafka, 1887 (type)
P. nephropiformis (Schlüter, 1862)
P. westfalica Mertin, 1941
Pseudohomarus†† van Hoepen, 1962 (0, 0, 1)
P. umsinenensis van Hoepen, 1962 (type)
Wongastacia†† Hu, 1984a (0, 0, 1)
W. taiwanica Hu, 1984a (type)
Thaumastochelidae† Bate, 1888
Oncopareia†† Bosquet, 1854 (=*Nymphaeops* Schlüter, 1862; =*Ischnodactylus* Pelseneer, 1886c; =*Stenocheles* Fritsch & Kafka, 1887) (0, 0, 17)
O. belgicus (Forir, 1887)
O. biseralis (Fritsch & Kafka, 1887)
O. bredai Bosquet, 1854 (type) (=*Hoploparia beyrichi* Schlüter, 1862; *Hoploparia bosqueti* Pelseneer, 1886c)
O. coesfeldiensis (Schlüter, 1862) (=*Enoploclytia paucispina* Schlüter *in* Schlüter & Von der Marck, 1868)
O. cookei (Rathbun, 1935a)
O. cultellus (Rathbun, 1935a)
O. dentatus (Rathbun, 1935a)
O. esocinus (Fritsch & Kafka, 1887)

O. *inaequidens* (Pelseneer, 1886c)
O. *klintebjergensis* Jakobsen & Collins, 1979
O. *lunatus* (Fritsch & Kafka, 1887)
O. *macrodactylus* (Schlüter, 1862)
O. *muncki* (Pelseneer, 1886c)
O. *parvulus* (Fritsch & Kafka, 1887)
O. *pectiniformis* (Böhm, 1891)
O. *sendenhorstensis* (Schlüter, 1862)
O. *texanus* (Rathbun, 1935a)
Superfamily Astacoidea Latreille, 1802
Family **Astacidae**† Latreille, 1802
Astacus† Fabricius, 1775 (3, 0, 4)
A. *edwardsi*†† Van Straelen, 1928b
A. *laevissimus*†† Fritsch & Kafka, 1887
A. *multicavatus*†† Bell, 1863
A. *pallipes*†† Lereboullet, 1858
Austropotamobius† Skorikov, 1907 (3, 0, 1)
A. *llopisi*†† (Vía, 1971)
Pacifastacus† Bott, 1950 (5, 0, 1)
P. *chenoderma*†† (Cope, 1871)
Cambaridae† Hobbs, 1942
Palaeocambarus†† Taylor, Schram & Yan-Bin, 1999 (0, 0, 1)
P. *licenti*†† (Van Straelen, 1928c) (type) (=*Astacus spinirostrius* Imiazumi, 1938)
Procambarus† Ortmann, 1905 (165, 0, 1)
P. *primaevus*†† (Packard, 1880)
Family **Cricoidoscelosidae**†† Taylor, Schram & Yan-Bin, 1999
Cricoidoscelosus†† Taylor, Schram & Yan-Bin, 1999 (0, 0, 1)
C. *aethus* Taylor, Schram & Yan-Bin, 1999 (type)
Superfamily Parastacoidea Huxley, 1879
Family **Parastacidae**† Huxley, 1879
Astacopsis† Huxley, 1879 (2, 1, 0)
A. *franklini*† (Gray, 1845)
Lammuastacus†† Aguirre-Urreta, 1992 (0, 0, 1)
L. *longirostris* Aguirre-Urreta, 1992 (type)
Palaeoechinastacus†† Martin et al., 2008 (0, 0, 1)

P. australianus Martin et al., 2008 (type)
Paranephrops† White, 1842 (2, 0, 1)
P. fordycei†† Feldmann & Pole, 1994
Incertae sedis
"*Enoploclytia*" *porteri* Miller & Ash, 1988
Infraorder THALASSINIDEA Latreille, 1831
Superfamily Thalassinoidea Latreille, 1831
Family **Thalassinidae**† Latreille, 1831
Thalassina† Latreille, 1806 (3, 1, 1)
T. anomala† (Herbst, 1804) (=*T. antiqua* Bell, 1844; *T. emeryi* Bell, 1845)
T. grandidactylus†† Robineau-Desvoidy, 1849
Superfamily Callianassoidea Dana, 1852a
Family **Axianassidae** Schmitt, 1924
No fossil representatives
Callianassidae† Dana, 1852a
Brecanclawu†† Schweitzer & Feldmann, 2001a (0, 0, 1)
B. rathbunae Schweitzer & Feldmann, 2001a (type)
Callianassa† Leach, 1814 (57, 0, 154) (=*Semiranina*†† Bachmayer, 1954a)
C. abbreviata†† Rathbun, 1918
C. affinis† A. Milne-Edwards, 1860
C. alabamensis†† Rathbun, 1935a
C. alaica†† Birshtein, 1956
C. almerai†† Müller, 1993
C. alpha†† Stenzel, 1935
C. americana†† Woods, 1922
C. anguillensis†† Rathbun, 1919
C. armata† A. Milne-Edwards, 1870
C. articulata† Rathbun, 1906
C. atlantica† Rathbun, 1926b
C. atrox†† Bittner, 1893
C. awakina†† Glaessner, 1960
C. bakeri†† Glaessner, 1948
C. bandonensis†† Rathbun, 1926b
C. batei†† Woodward, 1869
C. beberibae†† Beurlen, 1962
C. berryi†† Rathbun, 1935a

C. beta†† Stenzel, 1935
C. birmanica†† Noetling, 1901
C. bosqueana†† Rathbun, 1935a
C. brazoensis†† Stenzel, 1934
C. brevimanus†† Beurlen, 1939
C. brocchii†† Lőrenthey, 1898a
C. burlesonensis†† Stenzel, 1935
C. burckhardti†† Böhm, 1911
C. calaritana†† Ristori, 1896
C. canavarii†† Ristori, 1889a
C. chalmasii†† Brocchi, 1883
C. chanabadica†† Birshtein, 1956
C. cheyennensis†† Rathbun, 1930a
C. clarki†† Pilsbry, 1916
C. congolae†† Secretan, 1970
C. conradi†† Pilsbry, 1901
C. cowlitzensis†† Rathbun, 1926b
C. craterifera†† Lőrenthey *in* Lőrenthey & Beurlen, 1929
C. creagrachir†† (Schafhäutl, 1863)
C. cretacea†† Rathbun, 1935a
C. danai†† Hall & Meek, 1855
C. delta†† Rathbun, 1935a
C. depressiva†† Secretan, 1970
C. desmarestiana†† A. Milne-Edwards, 1860
C. edwardsi†† Vía, 1959
C. elegans†† Böhm, 1922
C. epsilon†† Rathbun, 1935a
C. erecta†† Böhm, 1926
C. espichelensis†† Veiga Ferreira, 1961
C. eta†† Rathbun, 1936
C. ferox†† Bittner, 1893
C. floriana†† Glaessner, 1928
C. floridana†† Rathbun, 1935a
C. fortis†† Böhm, 1922
C. fraasi†† Noetling, 1885b
C. frangens†† Böhm, 1922
C. fresnoensis†† Rathbun, 1926b

C. gamma†† Stenzel, 1935
C. gigantea†† Withers, 1924a
C. glabra†† Lőrenthey *in* Lőrenthey & Beurlen, 1929
C. heberti†† A. Milne-Edwards, 1860
C. hilli†† Rathbun, 1918
C. hulli†† Rathbun, 1935a
C. infracretacea†† Tribolet, 1874
C. incerta†† Rathbun, 1926b
C. inglisestris†† Roberts, 1953
C. inornata†† Nagao & Huzioka, 1938
C. isikariensis†† Nagao & Ôtatume, 1938
C. jahringensis†† Glaessner, 1928
C. kerepesiensis†† Müller, 1976
C. kewana†† Rathbun, 1926b
C. knapptonensis†† Rathbun, 1926b
C. kraussei†† Böhm, 1911 (=*Callianassa prisca* Krause, 1891)
C. kraussi† Stebbing *in* Kensley, 1985
C. lacunosa†† Rathbun, 1918
C. latidigitata†† Rathbun, 1919
C. ledae†† Fritsch, 1871
C. longa†† Noetling, 1885b
C. lusitanica†† Veiga Ferreira, 1954
C. macrodactylus†† A. Milne-Edwards, 1860
C. magna†† Rathbun, 1918
C. masanorii†† Karasawa, 1998a
C. massarandubae†† Beurlen, 1962
C. maxima†† A. Milne-Edwards, 1860
C. megachir†† (Schafhäutl, 1863)
C. menziesi†† Withers, 1926
C. meridionalis†† Ball, 1960
C. michelotti†† A. Milne-Edwards, 1860
C. minor†† Fischer-Benzon, 1866
C. moinensis†† Rathbun, 1918
C. mokkatamensis†† Noetling, 1885b
C. mottai†† Beurlen, 1962
C. munieri†† Brocchi, 1883
C. neocomiensis†† Woodward, 1869

C. nilotica†† Fraas, 1867
C. norica†† Glaessner, 1928
C. nuda†† Beurlen, 1939
C. oktibbehana†† Rathbun, 1935a
C. oregonensis†† Dana, 1849
C. orientalis† A. Milne-Edwards, 1860
C. oroszyi†† (Bachmayer, 1954a)
C. orthodactylus†† (Schafhäutl, 1863)
C. ovalis†† Rathbun, 1918
C. panamensis†† Glaessner, 1929b
C. parinasensis†† Woods, 1922
C. parisiensis†† A. Milne-Edwards, 1870
C. pedemontana†† Crema, 1895
C. pellucida†† Rathbun, 1919
C. peruviana†† Rathbun, 1947
C. primaeva†† Philippi, 1887
C. prisca†† A. Milne-Edwards, 1860
C. pseudofraasi†† Lőrenthey *in* Lőrenthey & Beurlen, 1929
C. pseudonilotica†† Lőrenthey *in* Lőrenthey & Beurlen, 1929
C. pseudorakosiensis†† Lőrenthey *in* Lőrenthey & Beurlen, 1929
C. pugnax†† Böhm, 1922
C. pustulata†† Withers, 1926
C. quadrata†† Rathbun, 1918
C. rakosiensis†† Lőrenthey, 1898a
C. rapax†† Bittner, 1893
C. rovasendae†† Crema, 1895
C. roztoczensis†† Müller, 1996
C. seefriedi†† Ammon, 1905
C. simplex†† Bittner, 1893
C. sismondai†† A. Milne-Edwards, 1860
C. songoensis†† Böhm, 1922
C. spinosa†† Tribolet, 1876
C. spinulosa†† Rathbun, 1918
C. stephensi†† Rathbun, 1926b
C. stimpsoni† Gabb, 1864

C. stridens†† Rathbun, 1918
C. subplana†† Withers, 1924a
C. subspinosa†† Glaessner, 1929b
C. subterranea† (Montagu, 1808)
C. suffolkensis†† Rathbun, 1935a
C. szobensis†† Müller, 1984a
C. taiwanica†† Hu & Tao, 1996
C. tanakai†† Imaizumi, 1957a
C. tayamai†† Imaizumi, 1952a
C. tenuis†† Rathbun, 1918
C. tepetatensis†† Rathbun, 1930b
C. theta†† Rathbun, 1936
C. transversoplicata†† Noetling, 1885b
C. trechmanni†† Withers, 1924a
C. turkestanica†† Birshtein, 1956
C. ulrichi†† White, 1880
C. umpquaensis†† Rathbun, 1926b
C. uncifera†† Harbort, 1905
C. urgoneensis†† Lőrenthey *in* Lőrenthey & Beurlen, 1929
C. valida†† Rathbun, 1935a
C. velox†† Bittner, 1893
C. vidali†† Vía, 1959
C. vorax†† Bittner, 1893
C. waikurana†† Glaessner, 1960
C. wechesensis†† Stenzel, 1934
C. yagii†† Imaizumi, 1957a
C. zeta†† Rathbun, 1936

Calliax† de Saint Laurent, 1973 (3, 0, 1)
C. whiteavesi†† (Woodward, 1896)

Callichirus† Stimpson, 1866 (3, 2, 3)
C. islagrande† (Schmitt, 1935)
C. major† (Say, 1818)
C. pustulosa†† (Withers, 1926)
C. symmetricus†† (Feldmann & Zinsmeister, 1984a)
C. waagei†† Crawford et al., 2006

Comoxianassa†† Schweitzer, Feldmann, Ćosović, Ross & Waugh, 2009 (0, 0, 1)

C. haggarti Schweitzer et al., 2009 (type)
Corallianassa† Manning, 1987 (12, 0, 2)
C. acucurvata†† Swen et al., 2001
C. rigoi†† De Angeli & Garassino, 2006a
Cowichianassa†† Schweitzer, Feldmann, Ćosović, Ross & Waugh, 2009 (0, 0, 1)
C. meckerti Schweitzer et al., 2009 (type)
Eucalliax† Manning & Felder, 1991 (8, 0, 4)
E. miyazakiensis†† Karasawa, 1993
E. vicetina†† Beschin, Busulini, De Angeli & Tessier, 2002
E. yatsuoensis†† (Karasawa, 1993)
E. yoshiroi†† Karasawa, 1992a
Eoglypturus†† Beschin, De Angeli, Checchi & Zarantello, 2005 (0, 0, 1)
E. grolensis Beschin et al., 2005 (type)
Glypturus† Stimpson, 1866 (3, 1, 1)
G. acanthochirus† Stimpson, 1866
G. toulai†† (Rathbun, 1919)
Lepidophthalmus Holmes, 1904 (13, 1, 0)
L. jamaicense ?† (Schmitt, 1935)
Melipal†† Schweitzer, Feldmann, Encinas & Suárez, 2006 (0, 0, 1)
M. chilensis Schweitzer et al., 2006 (type)
Neocallichirus† Sakai, 1988 (25, 0, 18)
N. aetodes†† Schweitzer, Iturralde-Vinent et al., 2006
N. allegranzii†† Beschin et al., 2005
N. bona†† (Imaizumi, 1959a)
N. borensis†† Beschin et al., 2006
N. dijki†† (Martin, 1883)
N. fortisi†† Beschin et al., 2002
N. manningi†† Schweitzer, Feldmann et al., 2003
N. matsoni†† (Rathbun, 1935a)
N. nishikawai†† (Karasawa, 1993)
N. okamotoi†† (Karasawa, 1993)
N. peraensis†† Collins et al., 1996
N. porterensis†† (Rathbun, 1926b)
N. quisquellanus†† Schweitzer, Iturralde-Vinent et al., 2006

N. rhinos†† Schweitzer & Feldmann, 2002

N. sakiae†† Karasawa & Fudouji, 2000

N. scotti†† (Brown & Pilsbry, 1913) (= *Callianassa crassa* A. Milne-Edwards, 1870; *C. crassimana* Rathbun, 1918; *C. miocenica* Rathbun, 1919)

N. vaughani†† (Rathbun, 1918)

N. wellsi†† Schweitzer, Feldmann & Gingerich, 2004

Neotrypaea† Manning & Felder, 1991 (3, 1, 0)

Neotrypaea gigas (Dana, 1852a)† (= *Callianassa longimana* Stimpson, 1857)

Podocallichirus† Sakai, 1999 (7, 0, 1)

P. grandis†† (Karasawa & Goda, 1996)

Protocallianassa†† Beurlen, 1930a (= *Mesostylus* Bronn & Roemer, 1852) (0, 0, 21)

P. antiqua (Roemer, 1841)

P. archiaci (A. Milne-Edwards, 1860) (type)

P. australica Glaessner, 1956

P. bohemica (Fritsch, 1867)

P. brevis (Fritsch, 1867)

P. caucasica Aliev & Aliev, 1980

P. cenomanensis (A. Milne-Edwards, 1860)

P. cliffwoodensis Roberts, 1962

P. elongata (Fritsch, 1867)

P. faujasi (Desmarest, 1822)

P. gracilis (Fritsch, 1867)

P. klofi Bishop, 1983a

P. mortoni (Pilsbry, 1901)

P. patagonica Aguirre Urreta, 1982

P. pilsbryi (Rathbun, 1935a)

P. pleuralum Beikirch & Feldmann, 1980

P. praecepta Roberts, 1962

P. russelli Bishop, 1985a

P. saetosa (Förster & Stinnesbeck, 1987)

P. sarysuensis Aliev & Aliev, 1980

P. tourtiae (Fritsch, 1867)

Sergio† Manning & Lemaitre, 1994 (5, 1, 0)

S. trilobatus† (Biffar, 1970)

Trypaea† Dana, 1852a (1, 0, 1)

T. mizunamiensis†† Karasawa, 1993

Vegarthron†† Schweitzer & Feldmann, 2002 (0, 0, 1)

V. santiago Schweitzer & Feldmann, 2002 (type)

Callianideidae Kossmann, 1880

No fossil representatives

Ctenochelidae† Manning & Felder, 1991

Callianopsis† de Saint Laurent, 1973 (2, 0, 8)

C. australis†† Casadío et al., 2004

C. clallamensis†† (Withers, 1924b)

C. elongatodigitata†† (Nagao, 1941)

C. inornatus†† Schweitzer & Feldmann, 2001a

C. kusiroensis†† (Nagao, 1941)

C. microspineus†† Beschin et al., 2005

C. muratai†† (Nagao, 1941)

C. titaensis†† (Nagao, 1941) (=*Callianassa shikamai* Imaizumi, 1957a)

Ctenocheles† Kishinouye, 1926 (=*Ischnodactylus* Pelseneer, 1886c) (5, 1, 15)

C. anderseni†† Collins & Jakobsen, 2003

C. chattiensis†† Polkowsky, 2004

C. cookei†† (Rathbun, 1935a)

C. cultellus†† (Rathbun, 1935a)

C. dentatus†† (Rathbun, 1935a)

C. falciformis†† Collins & Todd *in* Todd & Collins, 2005

C. hokoensis†† Schweitzer & Feldmann, 2001a

C. inaequidens†† (Pelseneer, 1886c)

C. madagascariensis†† Secretan, 1964

C. maorianus† Powell, 1949

C. ornatus†† Beschin et al., 2005

C. possagnoensis Busulini & Beschin, 2009

C. rupeliensis†† (Beurlen, 1939)

C. secretanae†† Schweitzer & Feldmann, 2002

C. sujakui†† Imaizumi, 1958

C. valdellae†† (Fabiani, 1908)

C. victor†† Glaessner, 1948

Gourretiidae† Sakai, 1999

Gourretia† de Saint Laurent, 1973 (10, 0, 1)

G. aquilae†† (Rathbun, 1935a)

Laomediidae† Borradaile, 1903
Jaxea† Nardo, 1847 (1, 1, 1)
J. kuemeli†† Bachmayer, 1954a
J. nocturna† Nardo, 1847
Laomedia† De Haan, 1847 (4, 0, 1)
L. praestacina†† Karasawa, 1989
Laurentiella† LeLoeuff & Intès, 1974 (1, 0, 1)
L. imaizumii†† Karasawa, 1993
Thomassiniidae de Saint Laurent, 1979
No fossil representatives
Upogebiidae† Borradaile, 1903
Upogebia† Leach, 1814 (114, 2, 21)
U. affinis† (Say, 1818)
U. barti†† Fraaije, Van Bakel et al., 2006
U. boehmi†† Glaessner, 1930
U. ? *clypeata*†† (Carter, 1898)
U. controversa†† (Tribolet, 1874)
U. digitata†† (Robineau-Desvoidy, 1849)
U. eocenica†† Rathbun, 1926b
U. gamma†† (Rathbun, 1935a)
U. kowai†† Feldmann et al., 2008
U. lambrechtsi†† Fraaije, Van Bakel et al., 2006
U. meyeri†† (Robineau-Desvoidy, 1849)
U. midwayensis†† Rathbun, 1935a
U. miocenica†† Garassino, Artal & Pasini, 2009c
U. mizunamiensis†† Karasawa, 1989
U. munsteri†† (Robineau-Desvoidy, 1849)
U. obscura†† (Von Meyer, 1834)
U. perarolensis†† De Angeli & Messina, 1992
U. rhacheochir†† Stenzel, 1945
U. scabra†† Müller, 1974a
U. stellata†† (Montagu, 1808)
U. striata†† Karasawa & Kishimoto, 1996
U. tanegashimensis†† Karasawa & Inoue, 1992
Upogebia sp. cf. *U. imperfecta*† Sakai, 1982
Superfamily Axioidea Huxley, 1879
Family **Axiidae**† Huxley, 1879
Axiopsis† Borradaile, 1903 (10, 0, 1)

A. eximia†† Kensley & Williams, 1990
Axius† Leach, 1816 [imprint 1815] (3, 0, 2)
A. fontannesi†† A. Milne-Edwards *in* Fontannes, 1885
A. reticulatus†† Rathbun, 1918
Etallonia†† Oppel, 1861b (0, 0, 4)
E. hoellorum Polz, 1999
E. longimana (Münster, 1839) (type)
E. subtilis (Krause, 1908)
E. suprajurensis (Quenstedt, 1852)
Huxleycaris†† Bravi & Garassino, 1998a (0, 0, 1)
H. beneventana Bravi & Garassino, 1998a (type)
Magila†† Münster, 1839 (0, 0, 12)
M. bonjouri (Étallon, 1861)
*M. denticulata (*Münster, 1839)
M. desmarestii (Münster, 1839)
M. dissimilis Carter, 1886
M. dura (Moericke, 1889)
M. latimana (Münster, 1839) (type)
M. levimana Carter, 1886
M. parvula Oppel, 1862
M. pichleri Oppel, 1862
M. robusta Oppel, 1862
M. rougemonti Tribolet, 1873
M. straeleni (Patrulius, 1959)
Megachela†† Schweigert, 2003 (0, 0, 1)
M. frickhingeri Schweigert, 2003 (type)
Protaxius†† Beurlen, 1930a (0, 0, 4)
P. betajurensis (Quenstedt, 1858)
P. eocenicus Secretan, 1975
P. isochelus (Woodward, 1876a) (type)
P. suevicus Beurlen, 1930a
Schlueteria†† Fritsch & Kafka, 1887 (0, 0, 4)
S. carinata Taylor, 1979
S. menabensis Secretan, 1964
S. tetracheles Fritsch & Kafka, 1887 (type)
S. tuberculosa Secretan, 1964
Calocarididae Ortmann, 1891
No fossil representatives

Micheleidae† Sakai, 1992
Paki†† Karasawa & Hayakawa, 2000 (0, 0, 1)
P. rurkonsimpu Karasawa & Hayakawa, 2000 (type)
Strahlaxiidae Poore, 1994
No fossil representatives
Infraorder PALINURA Latreille, 1802
Superfamily Eryonoidea De Haan, 1841
Family **Eryonidae**†† De Haan, 1841
Cycleryon†† Glaessner, 1965 (0, 0, 4)
C. elongatus (Münster, 1839)
C. orbiculatus (Münster, 1839) (=*E. armatus* Von Knebel, 1907; *E. subrotundus* Münster, 1839)
C. propinquus (Schlotheim, 1822) (=*E. schuberti* Von Meyer, 1836; *E. spinimanus* Germar, 1827) (type)
C. wulfi Garassino & Schweigert, 2004
Eryon†† Desmarest, 1822 (0, 0, 5)
E. arctiformis Schlotheim, 1822 (type)
E. ellipticus Van Straelen, 1923a
E. neocomiensis Woodward, 1881
E. perroni Étallon, 1859
E. sublevis Carter, 1886
Knebelia†† Van Straelen, 1922 (0, 0, 1)
K. bilobata (Münster, 1839) (type)
Rosenfeldia†† Garassino, Teruzzi & Dalla Vecchia, 1996 (0, 0, 2)
R. oppeli (Woodward, 1866b)
R. triasica Garassino et al., 1996 (type)
Coleiidae†† Van Straelen, 1924 [imprint 1925]
Coleia†† Broderip, 1835 (=*Palaeopolycheles*†† Von Knebel, 1907) (0, 0, 21)
C. antiqua Broderip, 1835 (type)
C. barrovensis (McCoy, 1849)
C. boboi Garassino & Gironi, 2006
C. bredonensis Woods, 1925
C. brodiei (Woodward, 1866b)
C. crassichelis (Woodward, 1866b) (=*Archaeoastacus (Eryon) willemoesii* Bate, 1884)
C. edwardsi (Morière, 1864)

C. incerta Secretan, 1964

C. longipes (Fraas, 1855)

C. mediterranea Pinna, 1968

C. moorei (Woodward, 1866b)

C. morierei (Renault, 1889)

C. pinnai Teruzzi, 1990

C. popeyei Teruzzi, 1990

C. sibirica Chernyschev, 1930

C. sinuata Beurlen, 1928a

C. tenuichelis Woods, 1925

C. theodorii Kuhn, 1952

C. uzume Karasawa et al., 2003

C. viallii Pinna, 1968

C. wilmcotensis (Woodward, 1866b)

**Hasaracancer*†† Jux, 1971 (0, 0, 1)

H. cristatus Jux, 1971 (type)

Hellerocaris†† Van Straelen, 1924 [imprint 1925] (0, 0, 1)

H. falloti (Van Straelen, 1923a) (type)

Proeryon†† Beurlen, 1928a (0, 0, 5)

P. giganteus (Van Straelen, 1923a)

P. hartmanni (Von Meyer, 1835) (type) [= *P. banzensis* Kuhn, 1952; *P. longiceps* Beurlen, 1928a; *P. macrophthalmus* (Krause, 1891); *P. richardsoni* (Woodward, 1911)]

P. laticaudatus Beurlen, 1928a (= *P. hauffi* Beurlen, 1944; *P. giganteus* Beurlen, 1930b)

P. stoddarti (Woodward, 1881)

P. viluensis Chernyshev, 1930

Pseudocoleia†† Garassino & Teruzzi, 1993 (0, 0, 1)

P. mazzolenii Garassino & Teruzzi, 1993 (type)

Tropifer†† Gould, 1857 (0, 0, 1)

T. laevis Gould, 1857 (type)

Palaeopentachelidae†† Ahyong, 2009

Palaeopentacheles†† Von Knebel, 1907 (0, 0, 2)

P. roettenbacheri (Münster, 1839) (type)

P. starri Schweitzer & Feldmann, 2001a

Polychelidae† Wood-Mason, 1875

Antarcticheles†† Aguirre-Urreta et al., 1990 (0, 0, 1)

A. antarcticus Aguirre-Urreta et al., 1990 (type)

Willemoesiocaris†† Van Straelen, 1924 [imprint 1925] (0, 0, 1)

W. ovalis (Van Straelen, 1923a) (type)

Incertae sedis

"*Coleia*" *escheri* (Oppel, 1862)

Palaeopolycheles crymensis Levitski, 1974

Tetrachelidae†† Beurlen, 1930a

Tetrachela†† Reuss, 1858 (0, 0, 1)

T. raiblana Bronn, 1858 (type)

Superfamily Palinuroidea Latreille, 1802

Family **Cancrinidae**†† Beurlen, 1930a

Cancrinos†† Münster, 1839 (0, 0, 3)

C. claviger Münster, 1839 (type) (= *C. latipes* Münster, 1839)

C. libanensis Garassino & Schweigert, 2006a

C. sahelalmae (Roger, 1944)

Palinuridae† Latreille, 1802

Archaeocarabus†† McCoy, 1849 (0, 0, 4)

A. bowerbanki McCoy, 1849 (type)

A. gardnerae Rathbun, 1935a

A. vanuaensis Rathbun, 1945

A. whitfieldi (Pilsbry, 1901)

Archaeopalinurus†† Pinna, 1974 (0, 0, 1)

A. levis Pinna, 1974 (type)

Astacodes†† Bell, 1863 (0, 0, 6)

A. davisi Stenzel, 1945

A. falcifer Bell, 1863 (type)

A. kamptneri (Bachmayer, 1959)

A. maxwelli Stenzel, 1945

A. strambergensis (Bachmayer, 1959)

A. woodwardi Fritsch & Kafka, 1887

Jasus† Parker, 1883 (6, 0, 1) (number of extant species from Booth, 2006)

J. flemingi†† Glaessner, 1960

Justitia† Holthuis, 1946 (1, 1, 1)

J. desmaresti† (Massalongo, 1854) (type)

J. vicetina†† Beschin, De Angelo & Garassino, 2001

Linuparus† White, 1847a (3, 0, 32)
- *L. adkinsi*†† Rathbun, 1935a
- *L. africanus*†† Glaessner, 1932a
- *L. bererensis*†† Secretan, 1964
- *L. bigranulatus*†† Glaessner, 1930
- *L. canadensis*†† (Whiteaves, 1885) (=*L. atavus* Ortmann, 1897)
- *L. carteri*†† (Reed, 1911)
- *L. dentatus*†† Van Straelen, 1936a
- *L. duelmenensis*†† (Geinitz, 1849-50)
- *L. dzheirantuiensis*†† Feldmann, Schweitzer et al., 2007
- *L. eocenicus*†† Woods, 1925
- *L euthymei*†† (Roman & Mazeran, 1920)
- *L. grimmeri*†† Stenzel, 1945
- *L. hantscheli*†† Mertin, 1941
- *L. japonicus*†† Nagao, 1931
- *L. kleinfelderi*†† Rathbun, 1931a
- *L. korura*†† Feldmann & Bearlin, 1988
- *L. laevicephalus*†† Mertin, 1941
- *L. macellarii*†† Feldmann & Tshudy, 1988
- *L. petkovici*†† Bachmayer & Marković, 1955
- *L. pustulosus*†† Feldmann, Bishop & Kammer, 1977
- *L. richardsi*†† Roberts, 1962
- *L. schluteri*†† (Tribolet, 1874)
- *L. scyllariformis*†† (Bell, 1858)
- *L. spinosus*†† Collins & Rasmussen, 1992
- *L. stolleyi*†† (Haas, 1889)
- *L. straili*†† (Forir, 1887)
- *L. tarrantensis*†† Davidson, 1963
- *L. texanus*†† Rathbun, 1935a
- *L. trigonus*† (Von Siebold, 1824)
- *L. vancouverensis*†† (Whiteaves, 1895)
- *L. watkinsi*†† Stenzel, 1945
- *L. wilcoxensis*†† Rathbun, 1935a

Palaeopalinurus†† Bachmayer, 1954b (0, 0, 1)
- *P. glaessneri* Bachmayer, 1954b (type)

Palinurina†† Münster, 1839 (0, 0, 2)
- *P. longipes* Münster, 1839 (type) (=*P. pygmea* Münster, 1839)

P. tenera Oppel, 1862

Palinurus† Weber, 1795 (5, 0, 3) (number of extant species from Groeneveld et al., 2006)

P. baumbergicus†† Schlüter, 1862

P. palaciosi†† Vega, García-Barrera et al., 2006

P. woodwardi†† Fritsch & Kafka, 1887

Panulirus† White, 1847a (21, 0, 1) (number of extant species from Patek et al., 2006)

P. destombesi†† Garassino & Breton, 2009

Pehuenchia†† Rusconi, 1948 (0, 0, 2)

P. ? *magna* Rusconi, 1948

P. tellecheai Rusconi, 1948 (type)

Scyllaridae† Latreille, 1825

Biarctus† Holthuis, 2002 (3, 1, 0)

B. vitiensis† (Dana, 1852b)

Palibacus†† Förster, 1984 (0, 0, 1)

P. praecursor (Dames, 1886) (type)

Parribacus† Dana, 1852a (5, 1, 3)

P. caesius†† Squires, 2001

P. cottreaui†† (Roger, 1946)

P. cristatus†† Förster, 1984

P. japonicus† Holthuis, 1960

Scyllarella†† Rathbun, 1935a (0, 0, 3)

S. aspera†† Rathbun, 1935a

S. gardneri†† (Woods, 1925)

S. gibbera†† Rathbun, 1935a (type)

Scyllarides† Gill, 1898 (13, 1, 4) (number of extant species from Spanier & Lavalli, 2006)

S. aequinoctialis† (Lund, 1793) (type)

S. bolcensis†† De Angeli & Garassino, 2008

S. koenigi†† (Bell, 1858)

S. punctatus†† Woods, 1925

S. tuberculatus†† (König, 1825)

Scyllarus† Fabricius, 1775 (9, 0, 2)

S. junghumi†† Böhm, 1922

S. mantelli†† Desmarest, 1822

Tricarinidae†† Feldmann, Kolahdouz, Biranvand & Schweigert, 2007

Tricarina†† Feldmann et al., 2007 (0, 0, 1)
T. gadvanensis Feldmann et al., 2007 (type)
Infraorder ANOMURA MacLeay, 1838
Superfamily Lomisoidea Bouvier, 1895
Family **Lomisidae** Bouvier, 1895
No fossil representatives
Superfamily Galatheoidea Samouelle, 1819
Family **Aeglidae**† Dana, 1852a
Haumuriaegla†† Feldmann, 1984 (0, 0, 1)
H. glaessneri Feldmann, 1984 (type)
Protaegla†† Feldmann, Vega, Applegate & Bishop, 1998 (0, 0, 1)
P. miniscula Feldmann et al., 1998 (type)
Chirostylidae† Ortmann, 1892a
Eumunida† Smith, 1883 (29, 0, 1)
E. pentacantha (Müller & Collins, 1991)
Pristinaspina†† Schweitzer & Feldmann, 2000a (0, 0, 1)
P. gelasina Schweitzer & Feldmann, 2000a (type)
Galatheidae† Samouelle, 1819
Acanthogalathea†† Müller & Collins, 1991 (0, 0, 3)
A. feldmanni De Angeli & Garassino, 2002
A. parva (Müller & Collins, 1991) (type)
A. squamosa Beschin, Busulini, De Angeli & Tessier, 2007
Austromunida†† Schweitzer & Feldmann, 2000a (0, 0, 1)
A. casadioi Schweitzer & Feldman, 2000a (type)
Brazilomunida†† Martins-Neto, 2001 (0, 0, 1)
B. brasiliensis (Beurlen, 1965) (type)
Calteagalathea†† De Angeli & Garassino, 2006a (0, 0, 1)
C. friulana De Angeli & Garassino, 2006a (type)
Cretagalathea†† Garassino, De Angeli & Pasini, 2008 (0, 0, 1)
C. exigua Garassino, De Angeli & Pasini, 2008 (type)
Eomunidopsis†† Vía, 1981 (0, 0, 7)
E. kinokunica Karasawa, Ohara & Kato, 2008
E. limonitica (Stenzel, 1945)
E. meerssensis Collins et al., 1995
E. neojurensis Patrulius, 1959

E. navarrensis (Van Straelen, 1940b)
E. orobensis (Ruiz de Gaona, 1943) (type)
E. portlandica Fraaye & Collins, 1996
Faxegalathea†† Jakobsen & Collins, 1997 (0, 0, 1)
F. platyspinosa Jakobsen & Collins, 1997 (type)
Galathea† Fabricius, 1793 (68, 2, 10)
G. affinis†† Ristori, 1886
G. antiqua†† Risso, 1815
G. berica†† De Angeli & Garassino, 2002
G. dispersa† Bate, 1859
G. keijii†† Karasawa, 1993
G. lupiae†† Robineau-Desvoidy, 1849
G. sahariana†† Garassino, De Angeli & Passini, 2008
G. spitzbergica†† Gripp, 1927
G. squamifera† Leach, 1815
G. strigifera†† Fischer-Benzon, 1866
G. valmaranensis†† De Angeli & Garassino, 2002
G. weinfurteri†† Bachmayer, 1950
Gastrosacus†† Von Meyer, 1851b (=*Galatheites* Balss, 1913) (0, 0, 11)
G. carteri Van Straelen, 1924 [imprint 1925]
G. eminens (Blaschke, 1911)
G. eutecta (Moericke, 1889)
G. meyeri (Moericke, 1889)
G. robineaui (Tribolet, 1874)
G. royoi Van Straelen, 1927
G. straeleni (Ruiz de Gaona, 1943)
G. tuberosa (Remeš, 1895)
G. tuberosiformis Lőrenthey *in* Lőrenthey & Beurlen, 1929
G. wetzleri Von Meyer, 1851b (type)
G. zitteli (Moericke, 1889)
Lessinigalathea†† De Angeli & Garassino, 2002 (0, 0, 1)
L. regale De Angeli & Garassino, 2002 (type)
Lophoraninella†† Glaessner, 1946 (0, 0, 1)
L. cretacea (Dames, 1886) (type)
Luisogalathea†† Karasawa & Hayakawa, 2000 (0, 0, 3)
L. cobbani (Bishop, 1985c)

L. cretacea (Stenzel, 1945)
L. tomitai Karasawa & Hayakawa, 2000 (type)
Mesogalathea†† Houša, 1963 (0, 0, 1)
M. striata (Remeš, 1895) (type)
Munida† Leach, 1820 (246, 0, 5)
M. konara†† Schweitzer & Feldmann, 2000a
M. nishioi†† Karasawa, 1993
M. ogaensis†† (Hatai & Kotaka, 1970)
M. primaeva†† Segerberg, 1900
M. quadroblonga†† Schweitzer & Feldmann, 2000a
Munidopsis† Whiteaves, 1874 (224, 0, 2)
M. foersteri Feldmann et al., 1993
M. scabrosa Feldmann & Wilson, 1988
Munitheites†† Lőrenthey *in* Lőrenthey & Beurlen, 1929 (0, 0, 1)
M. palfyi Lőrenthey *in* Lőrenthey & Beurlen, 1929 (type)
Paragalathea†† Patrulius, 1959 (0, 0, 11)
P. africana Garassino, De Angeli & Pasini, 2008
P. miyakoensis Takeda & Fujiyama, 1983
P. multisquamata Vía, 1981
P. neocomiensis (Van Straelen, 1936a)
P. ornatissima Patrulius, 1966
P. ruizi (Van Straelen, 1940b)
P. straeleni (Ruiz de Gaona, 1943)
P. striata (Remeš, 1895)
P. substriata (Blaschke, 1911)
P. ubaghsi (Pelseneer, 1886c)
P. verrucosa (Moericke, 1889) (type)
Palaeomunida†† Lőrenthey, 1902 (0, 0, 3)
P. defecta (Lőrenthey, 1902) (type)
P. moesica Muţiu & Bădăluţă, 1971
P. multicristata De Angeli & Garassino, 2002
Palaeomunidopsis†† Van Straelen, 1924 [imprint 1925] (0, 0, 1)
P. moutieri (Van Straelen, 1923b) (type)
Protomunida†† Beurlen, 1930a (0, 0, 2)
P. munidoides (Segerberg, 1900) (type)

P. pentacantha Müller & Collins, 1991
Shinkaia† Baba & Williams, 1998 (1, 0, 1)
S. katapsyxis†† Schweitzer & Feldmann, 2008a
Spathagalathea†† De Angeli & Garassino, 2002 (0, 0, 1)
S. minuta De Angeli & Garassino, 2002 (type)
Porcellanidae† Haworth, 1825
Annieporcellana†† Fraaije et al., 2008 (0, 0, 1)
A. dhondtae Fraaije et al., 2008 (type)
Beripetrolisthes†† De Angeli & Garassino, 2002 (0, 0, 1)
B. mulleri De Angeli & Garassino, 2002 (type)
Eopetrolisthes†† De Angeli & Garassino, 2002 (0, 0, 1)
E. striatissimus (Müller & Collins, 1991) (type)
Lobipetrolisthes†† De Angeli & Garassino, 2002 (0, 0, 1)
L. blowi De Angeli & Garassino, 2002 (type)
Longoporcellana†† Müller & Collins, 1991 (0, 0, 2)
L. denticulata Müller & Collins, 1991 (type)
L. lobata De Angeli & Garassino, 2002
Pachycheles† Stimpson, 1858a (39, 1, 2)
P. dorsosulcatus†† Beschin, Busulini, De Angeli & Tessier, 2007
P. latus†† Rathbun, 1918
P. stevensii† Stimpson, 1858a
Petrolisthes† Stimpson, 1858a (100, 0, 9)
P. avitus†† Rathbun, 1918
P. bittneri†† De Angeli & Garassino, 2002
P. haydni†† Müller, 1984a
P. lauensis†† Rathbun, 1945
P. magnus†† Müller, 1984a
P. miocaenicus†† Karasawa, 1993
P. myakkensis†† Bishop & Portell, 1989
P. ? striatissimus†† Müller & Collins, 1991
P. vicetinus†† Beschin, De Angeli & Checchi, 2001
Pisidia† Leach, 1820 (10, 0, 3)
P. dorsosinuosa†† De Angeli & Garassino, 2002
P. kokayi†† (Müller, 1974b)
P. viai†† Müller, 1984b
Polyonyx† Stimpson, 1858a (18, 0, 1)
P. arcuatus†† Müller & Collins, 1991

Porcellana† Lamarck, 1801 (10, 0, 1)
P. antiqua†† A. Milne-Edwards, 1862a
Retrorsichelidae†† Feldmann, Tshudy & Thomson, 1993
Retrorsichela†† Feldmann, Tshudy & Thomson, 1993 (0, 0, 1)
R. laevis Feldmann et al., 1993 (type)
Family uncertain
Ovocarcinus†† Müller & Collins, 1991 (0, 0, 1)
O. elongatus Müller & Collins, 1991 (type)
Superfamily Hippoidea Latreille, 1825
Family **Albuneidae**† Stimpson, 1858a
Albunea† Weber, 1795 (22, 0, 4)
A. asymmetrica†† (Müller, 1978)
A. cuisiana†† Beschin & De Angeli, 1984
A. hahnae†† Blow & Manning, 1996
A. turritellacola†† Fraaije, Van Bakel & Jagt, 2008
Italialbunea†† Boyko, 2002 (0, 0, 1)
I. lutetiana (Beschin & De Angeli, 1984) (type)
Praealbunea†† Fraaije, 2002 (0, 0, 1)
P. rickorum Fraaije, 2002 (type)
Stemonopa† Efford & Haig, 1968 (1, 0, 1)
S. prisca†† De Angeli, Beschin & Checchi, 2005
Zygopa† Holthuis, 1961 (2, 0, 1)
Z. galantensis†† (De Angeli & Marangon, 2001)
Blepharipodidae† Boyko, 2002
Lophomastix† Benedict, 1904 (3, 0, 4)
L. altoonaensis†† Nyborg & Vega, 2008
L. antiqua†† Schweitzer & Boyko, 2000
L. boykoi†† Nyborg & Vega, 2008
L. kellyi†† Nyborg & Vega, 2008
Hippidae Latreille, 1825
No fossil representatives
Superfamily Kiwaoidea Macpherson, Jones & Segonzac, 2005
Family **Kiwaidae** Macpherson, Jones & Segonzac, 2005
No fossil representatives
Superfamily Lithodoidea Samouelle, 1819
Family **Lithodidae**† Samouelle, 1819
Paralomis† White, 1856 (56, 0, 1)

P. debodeorum†† Feldmann, 1998
Superfamily Paguroidea Latreille, 1802
Family uncertain
Ammopylocheles†† Van Bakel et al., 2008 (0, 0, 1)
A. mclaughlini Van Bakel et al., 2008 (type)
Eopaguropsis†† Van Bakel et al., 2008 (0, 0, 1)
E. loercheri Van Bakel et al., 2008 (type)
Eotylaspis†† Van Bakel et al., 2008 (0, 0, 1)
E. wehnerae Van Bakel et al., 2008 (type)
Jurapylocheles†† Van Bakel et al., 2008 (0, 0, 1)
J. malutka Van Bakel et al., 2008 (type)
Family **Coenobitidae**† Dana, 1851c
Birgus† Leach, 1816 [imprint 1815] (1, 0, 1)
B. latro† (Linnaeus, 1767)
Coenobita† Latreille, 1829 (15, 1, 0)
C. rugosus† Bouvier, 1890 (= *C. granulata* Bouvier, 1890)
Diogenidae† Ortmann, 1892a
Annuntidiogenes†† Fraaije et al., 2008 (0, 0, 3)
A. ruizdegaonai Fraaije et al., 2008 (type)
A. sunuciorum Fraaije et al., 2008
A. worfi Fraaije et al., 2009
Calcinus† Dana, 1851a (43, 0, 1)
C. agnoensis†† Beschin et al., 2005
Ciliopagurus† Forest, 1995 (16, 0, 3)
C. obesus†† Van Bakel, Jagt & Fraaije, 2002
C. substriatiformis†† (Lőrenthey *in* Lőrenthey & Beurlen, 1929)
C. tethysianus†† Beschin, Busulini, De Angeli & Tessier, 2007
Dardanus† Paul'son, 1875 (40, 3, 8)
D. arnoldi†† Rathbun, 1926a
D. arrosor† (Herbst, 1796)
D. biordines†† Collins & Todd *in* Todd & Collins, 2005
D. curtimanus†† Müller & Collins, 1991
D. gemmatus† (H. Milne Edwards, 1848)
D. hungaricus†† (Lőrenthey *in* Lőrenthey & Beurlen, 1929)

D. impressus† (De Haan, 1849)

D. lauensis†† Rathbun, 1945

D. mediterraneus†† (Lőrenthey, 1907a [German edition 1909a])

D. muelleri†† Karasawa & Inoue, 1992

D. squamatus†† Collins *in* Collins et al., 2009

Diogenes† Dana, 1851a (56, 2, 2)

D. edwardsii† (De Haan, 1849)

D. longimanus†† Müller & Collins, 1991

D. matrensis†† Müller, 1984a

D. pugilator† (Roux, 1828)

Eocalcinus†† Vía, 1959 (0, 0, 2)

E. cavus Beschin et al., 2002

E. eocenicus Vía, 1959 (type)

Paguristes† Dana, 1851a (116, 0, 15)

P. baldoensis Garassino, De Angeli & Pasini, 2009b

P. chipolensis†† Rathbun, 1935a

P. cserhatensis†† Müller, 1984a

P. extentus†† Beschin, Busulini, De Angeli & Tessier, 2007

P. florae†† Collins et al., 1995

P. hokoensis†† Schweitzer & Feldmann, 2001a

P. johnsoni†† Rathbun, 1935a

P. lineatuberculatus†† Beschin et al., 2006

P. oligotuberculatus†† Müller & Collins, 1991

P. ouachitensis†† Rathbun, 1935a

P. prealpinus†† Beschin et al., 2005

P. santamartaensis†† Feldmann, Tshudy & Thomson, 1993

P. subaequalis†† (Rathbun, 1926a)

P. wheeleri†† Blow & Manning, 1996

P. whitteni†† Bishop, 1983c

Parapaguristes†† Bishop, 1986a (0, 0, 1)

P. tuberculatus Bishop, 1986a (type)

Petrochirus† Stimpson, 1858a (3, 2, 6)

P. bahamensis† (Herbst, 1791)

P. bouvieri†† Rathbun, 1918

P. diogenes† (Linnaeus, 1758)

P. inequalis†† Rathbun, 1919
P. mezi†† (Lőrenthey, 1907b [German edition 1909b])
P. poscolensis†† Beschin et al., 2006
P. priscus†† (Brocchi, 1883)
P. taylori†† Rathbun, 1935a
Striadiogenes†† Garassino, De Angeli & Pasini, 2009a (0, 0, 1)
S. frigerioi Garassino, De Angeli & Pasini, 2009a (type)

Paguridae† Latreille, 1802
Anapagurus† Henderson, 1886 (20, 0, 2)
A. carinatus†† Harvey, 1998 (for *A. marginatus* Müller, 1979)
A. miocenicus†† Müller, 1978
Diacanthurus† McLaughlin & Forest, 1997 (5, 0, 1)
D. clifdenensis†† (Hyden & Forest, 1980)
Goniochirus†† Étallon, 1861 (0, 0, 7)
G. antoinei Van Straelen, 1924 [imprint 1925]
G. babeaui Étallon, 1861 (type)
G. cristatus Carter, 1886
G. jaccardi Étallon, 1861
G. laevis Van Straelen, 1924 [imprint 1925]
G. obliquus Van Straelen, 1924 [imprint 1925]
G. platycheles (McCoy, 1849)
Orhomalus†† Étallon, 1861 (0, 0, 17)
O. araricus Étallon, 1861
O. astartinus Étallon, 1861
O. corallinus Étallon, 1861
O. deformis (Oppel, 1862)
O. deslongchampsi Hée, 1924
O. jaccardi (Tribolet, 1873)
O. macrochirus Étallon, 1861
O. magnificus Crônier & Courville, 2004
O. oppeli Étallon, 1861
O. oxfordensis (Hée, 1924)
O. pidauceti Étallon, 1861
O. portlandicus Étallon, 1861
O. rotulensis De Gregorio, 1884
O. spinosus Schweitzer, Feldmann & Lazăr, 2009

O. tombecki Tribolet, 1875
O. verrucosus Étallon, 1861
O. virgulinus Étallon, 1861 (type)
Pagurus† Fabricius, 1775 (169, 7, 19)
P. alabamensis†† Rathbun, 1935a
P. alatoides†† Philippe & Secretan, 1971
P. arrosor† (Herbst, 1794)
P. avellanedae†† Vía, 1951
P. banderensis†† Rathbun, 1935a
P. bernhardus† Linnaeus, 1758
P. concavus†† Müller, 1978
P. constans† (Stimpson, 1858a)
P. convexus†† Whetstone & Collins, 1982
P. dubius†† Noetling, 1885b
P. granosimanus† (Stimpson, 1862)
P. langei†† Collins & Jakobsen, 2003
P. latidactylus†† Müller & Collins, 1991
P. malloryi†† Schweitzer & Feldmann, 2001a
P. manzonii†† (Ristori, 1888)
P. marceti†† Vía, 1959
P. marini†† Vía, 1959
P. megalops† (Stimpson, 1858a)
P. pectinatus† (Stimpson, 1858a)
P. rakosensis†† Müller, 1978
P. similis† (Ortmann, 1892a)
P. squamosus†† Ristori, 1886
P. substriatus†† A. Milne-Edwards *in* Sismonda, 1861
P. travisensis†† Stenzel, 1945
P. tuberculosus†† Harvey, 1998 (for *P. albus* Müller, 1978)
P. turcus†† Müller, 1984a
Palaeopagurus†† Van Straelen, 1924 [imprint 1925] (0, 0, 14)
P. acutus Crônier & Courville, 2004
P. cretaceous Mertin, 1941
P. deslongchampsi Van Straelen, 1924 [imprint 1925] (type)
P. guilleaumei Van Straelen, 1924 [imprint 1925]

P. jurensis (Étallon, 1861)
P. kellowiensis (Étallon, 1861)
P. laevis Van Straelen, 1924 [imprint 1925]
P. morinicus Sauvage, 1891
P. neraudeaui Crônier & Courville, 2004
P. pilsbryi Roberts, 1962
P. quadratus Van Straelen, 1924 [imprint 1925]
P. serialis (Carter, 1886)
P. spinosus Van Straelen, 1924 [imprint 1925]
P. tuberculatus Van Straelen, 1924 [imprint 1925]
Pylopagurus† A. Milne-Edwards & Bouvier, 1891 (5, 0, 2)
P. corallinus†† Müller, 1996
P. leganyii†† Müller, 1984a
Parapaguridae Smith, 1882
No fossil representatives
Pylochelidae Bate, 1888
No fossil representatives
Pylojacquesidae McLaughlin & Lemaitre, 2001
No fossil representatives
Infraorder BRACHYURA Linnaeus, 1758
Section Dromiacea De Haan, 1833
Superfamily Eocarcinoidea Withers, 1932a
Family **Eocarcinidae**†† Withers, 1932a
Eocarcinus†† Withers, 1932a (0, 0, 1)
E. praecursor Withers, 1932a (type)
Superfamily Dakoticancroidea†† Rathbun, 1917a
Family **Ibericancridae**†† Artal et al., 2008
Ibericancer†† Artal et al., 2008 (0, 0, 1)
I. sanchoi Artal et al., 2008 (type)
Family **Dakoticancridae**†† Rathbun, 1917a
Avitelmessus†† Rathbun, 1923a (0, 0, 1)
A. grapsoideus Rathbun, 1923a (type)
Dakoticancer†† Rathbun, 1917a (0, 0, 2)
D. australis Rathbun, 1935a
D. overanus Rathbun, 1917a (type)
Seorsus†† Bishop, 1988a (0, 0, 1)
S. wadei Bishop, 1988a (type)

Tetracarcinus†† Weller, 1905 (0, 0, 1)
T. subquadratus Weller, 1905 (type)

Superfamily Homolodromioidea Alcock, 1900a

Family **Homolodromiidae**† Alcock, 1900a
Antarctidromia†† Förster et al., 1985 (0, 0, 1)
A. inflata Förster et al., 1985 (type)
Eoprosopon†† Förster, 1986 (0, 0, 1)
E. klugi Förster, 1986 (type)
Homolodromia† A. Milne-Edwards, 1880a (5, 0, 2)
H. chaneyi†† Feldmann & Wilson, 1988
H. novaezelandica†† Feldmann, 1993
Palehomola†† Rathbun, 1926a (0, 0, 1)
P. gorrelli Rathbun, 1926a (type)
Preclarocarcinus†† Schweitzer, Feldmann, Ćosović, Ross & Waugh, 2009 (0, 0, 1)
P. parvus Schweitzer, Feldmann, Ćosović, Ross & Waugh, 2009 (type)
Rhinodromia†† Schweitzer et al., 2004 (0, 0, 1)
R. richardsoni (Woodward, 1896) (type)

Bucculentidae†† Schweitzer & Feldmann, 2009a
Bucculentum†† Schweitzer & Feldmann, 2009a (0, 0, 2)
B. bachmayeri Schweitzer & Feldmann, 2009a
B. bucculentum (Wehner, 1988) (type)
Wilmingtonia†† Wright & Collins, 1972 (0, 0, 1)
W. satyrica Wright & Collins, 1972 (type)

Goniodromitidae†† Beurlen, 1932
Cycloprosopon†† Lőrenthey *in* Lőrenthey & Beurlen, 1929 (0, 0, 3)
C. complanatiforme (Moericke, 1889)
C. etalloni (Gemmellaro, 1869)
C. typicum (Lőrenthey *in* Lőrenthey & Beurlen, 1929) (type)
Cyclothyreus†† Remeš, 1895 (0, 0, 9)
C. cardiacus Schweitzer & Feldmann, 2009c
C. divaricatus Schweitzer & Feldmann, 2009c
C. latus (Moericke, 1889)
C. quadrophthalmus Schweitzer & Feldmann, 2009c
C. strangus Schweitzer & Feldmann, 2009c

C. transitorius (Remeš, 1895)
C. reussi (Gemmellaro, 1869)
C. strambergensis Remeš, 1895 (type)
C. tithonius (Gemmellaro, 1869)
Distefania†† Checchia-Rispoli, 1917 (=*Cyphonotus* [*sensu*] Bell, 1863) (0, 0, 10)
D. autissiodorensis (Van Straelen, 1936a)
D. centrosa (Van Straelen, 1940b)
D. cryptica (Jagt, Van Bakel & Fraaije, 2007)
D. himeraensis (Checchia-Rispoli, 1917) (=*D. sicula* Checchia-Rispoli, 1917) (type)
D. incerta (Bell, 1863)
D. oxythyreiformis (Gemmellaro, 1869)
D. renevieri Tribolet, 1876
D. sergipensis Beurlen, 1965
D. sinuososulcata Wright & Collins, 1972
D. transiens Wright & Collins, 1972
Eodromites†† Patrulius, 1959 (0, 0, 6)
E. grande (Von Meyer, 1857) (type)
E. depressus (Von Meyer, 1857)
E. dobrogea (Feldmann, Lazăr & Schweitzer, 2006)
E. nitidus (A. Milne-Edwards, 1865a)
E. polyphemi (Gemmellaro, 1869)
E. rostratus (Von Meyer, 1840a)
Goniodromites†† Reuss, 1858 [imprint 1857] (=*Iberihomola* Van Straelen, 1940b) (0, 0, 12)
G. aliquantulus Schweitzer, Feldmann & Lazăr, 2007
G. bidentatus Reuss, 1858 [imprint 1857] (type)
G. complanatus Reuss, 1858 [imprint 1857]
G. cenomanensis (Wright & Collins, 1972)
G. dacica (Von Mücke, 1915)
G. dentatus Lőrenthey *in* Lőrenthey & Beurlen, 1929
G. hirotai Karasawa & Kato, 2007
G. laevis (Van Straelen, 1940b)
G. polyodon Reuss, 1858 [imprint 1857]
G. sakawense Karasawa & Kato, 2007
G. serratus Beurlen, 1929
G. transsylvanicus Lőrenthey *in* Lőrenthey & Beurlen, 1929

Palaeodromites†† A. Milne-Edwards, 1865a (0, 0, 1)
P. octodentatus A. Milne-Edwards, 1865a (type)
Pithonoton†† Von Meyer, 1842 (0, 0, 7 *s.s.* and 6 *s.l.*) [meaning: 7 *sensu stricto* and 6 *sensu lato*]
P. elongatum (Von Meyer, 1860) *s.s.*
P. iyonofutanajima Karasawa & Kato, 2007 *s.s.*
P. laevimarginatum Lőrenthey *in* Lőrenthey & Beurlen, 1929 *s.s.*
P. marginatum Von Meyer, 1842 (type) *s.s.*
P. pustulosus (Von Meyer, 1860) *s.s.*
P. rotundum (Beurlen, 1925) *s.s.*
P. simplex (Von Meyer, 1837) *s.s.*
P. aequilatum (Von Meyer, 1860) *s.l.*
P. bourgeati Van Straelen, 1924 [imprint 1925] *s.l.*
P. hypocritum Étallon, 1861 *s.l.*
P. incisum Van Straelen, 1924 [imprint 1925] *s.l.*
P. moutieri (Hée, 1924) *s.l.*
P. obtusum (Von Meyer, 1860) *s.l.*
Plagiophthalmus†† Bell, 1863 (0, 0, 3)
P. collinsi Feldmann, Tshudy & Thomson, 1993
P. nitonensis Wright & Wright, 1950
P. oviformis Bell, 1863 (type)
Sabellidromites†† Schweitzer & Feldmann, 2008 [imprint 2007] (0, 0, 2)
S. inflata (Collins & Karasawa, 1993)
S. scarabaea (Wright & Wright, 1950) (type)
Trachynotocarcinus†† Wright & Collins, 1972 (0, 0, 2) (= *Trachynotus* Bell, 1863)
T. naglei (Bishop, 1983a)
T. sulcatus (Bell, 1863) (type)

Prosopidae†† Von Meyer, 1860
Laeviprosopon†† Glaessner, 1933a (0, 0, 7)
L. fraasi (Moericke, 1889)
L. grandicentrum Schweitzer & Feldmann, 2008b
L. icaunensis (Van Straelen, 1936a)
L. laeve Von Meyer, 1857 (type)
L. laculatum Schweitzer & Feldmann, 2008b
L. punctatum (Von Meyer, 1857)

L. sublaeve (Von Meyer, 1857)
Prosopon†† Von Meyer, 1835 (0, 0, 6)
P. abbreviatum Schweitzer & Feldmann, 2009a
P. aculeatum Von Meyer, 1857
P. mammillatum Woodward, 1868c
P. tuberosum Von Meyer, 1840a (type)
P. verrucosum Reuss, 1858 [imprint 1857]
P. ? dzhafarberdensis (Ilyin, 2005)
Protuberosa†† Schweitzer & Feldmann, 2009a (0, 0, 1)
P. protuberosa (Wehner, 1988) (type)
Tanidromitidae†† Schweitzer & Feldmann, 2008 [imprint 2007]
Gabriella†† Collins, Ross, Genzano & Mianzan, 2006 (= *Foersteria* Wehner, 1988) (0, 0, 5)
G. anfracta Schweitzer & Feldmann, 2009d
G. biburgensis (Wehner, 1988) (type)
G. lugobaensis (Förster, 1985)
G. patula Schweitzer & Feldmann, 2009d
G. plana (Van Straelen, 1936a)
Tanidromites†† Schweitzer & Feldmann, 2008 [imprint 2007] (0, 0, 4)
T. etalloni (Collins *in* Collins & Wierzbowski, 1985)
T. insignis (Von Meyer, 1857) (type)
T. richardsoni (Woodward, 1907)
T. sculpta (Quenstedt, 1857) (= *Prosopon lingulata* Von Meyer, 1858)
Incertae sedis
Mesodromilites†† Woodward, 1900b (0, 0, 1)
M. glaber (Woodward, 1898) (= *M. birleyae* Woodward, 1900b) (type)
Microcorystes†† Fritsch, 1893 (0, 0, 1)
M. parvulus Fritsch, 1893 (type)
Mithracites†† Gould, 1859 (0, 0, 1)
M. vectensis Gould, 1859 (type)
Nipponopon†† Karasawa, Kato & Terabe, 2006 (0, 0, 1)
N. hasegawai Karasawa, Kato & Terabe, 2006 (type)
Oonoton†† Glaessner, 1980 (0, 0, 1)
O. woodsi Glaessner, 1980 (type)

"*Oxythyreus*"
O. bjorki (Bishop & Williams, 2000)
O. izetti (Bishop, 1988b)
Rugafarius†† Bishop, 1985a (0, 0, 1)
R. fredrichi Bishop, 1985a (type)
Superfamily Glaessneropsoidea†† Patrulius, 1959
Glaessneropsidae†† Patrulius, 1959
Ekalakia†† Bishop, 1976 (0, 0, 2)
E. exophthalmops Feldmann, Schweitzer & Wahl, 2008
E. lamberti Bishop, 1976 (type)
Glaessneropsis†† Patrulius, 1959 (0, 0, 4)
G. bucegiana Patrulius, 1959
G. heraldica (Moericke, 1889) (type)
G. myrmekia Schweitzer & Feldmann, 2009a
G. tribulosa Schweitzer & Feldmann, 2009a
Oxythyreus [*sensu*] Reuss, 1858 [imprint 1857] (0, 0, 2)
O. gibbus (Reuss, 1858 [imprint 1857]) (type)
O. schneideri (Stolley, 1924)
Rathbunopon†† Stenzel, 1945 (0, 0, 5)
R. atherfieldensis Wright, 1997
R. obesum (Van Straelen, 1944)
R. oblitum (Carter, 1898)
R. polyakron Stenzel, 1945 (type)
R. woodsi Withers, 1951
Vectis†† Withers, 1946 (0, 0, 3)
V. caseyi Wright & Collins, 1972
V. echinorum Wright & Collins, 1972
V. wrighti Withers, 1946 (type)
Verrucarcinus†† Schweitzer & Feldmann, 2009a (0, 0, 2)
V. ordinatus (Collins *in* Collins & Wierzbowski, 1985)
V. torosus (Von Meyer, 1857) (type)
Lecythocaridae†† Schweitzer & Feldmann, 2009a
Lecythocaris†† Von Meyer, 1858 (0, 0, 2)
L. obesa Schweitzer & Feldmann, 2009a
L. paradoxa (Von Meyer, 1858) (type)
Longodromitidae†† Schweitzer & Feldmann, 2009a
Abyssophthalmus†† Schweitzer & Feldmann, 2009a (0, 0, 5)

A. hebes (Von Meyer, 1835)
A. mirus (Moericke, 1889)
A. schultzi Schweitzer & Feldmann, 2009b
A. spinosus (Von Meyer, 1842) (type)
A. stotzingensis (Von Meyer, 1856)
Coelopus†† Étallon, 1861 (0, 0, 4)
C. bigoti Hée, 1924
C. hoheneggeri (Moericke, 1889)
C. jolyi Étallon, 1861 (type)
C. tuberculatus Lőrenthey *in* Lőrenthey & Beurlen, 1929
Longodromites†† Patrulius, 1959 (0, 0, 5)
L. angustus (Reuss, 1858 [imprint 1857]) (type)
L. bicornutus Muţiu & Bădăluţă, 1971
L. excisus (Von Meyer, 1857)
L. meyeri (Étallon, 1859)
L. ovalis (Moericke, 1889)
Planoprosopon†† Schweitzer, Feldmann & Lazăr, 2007 (0, 0, 8)
P. aequus (Von Meyer, 1857)
P. cornutus (Wehner, 1988)
P. dumosus (Wehner, 1988)
P. heydeni (Von Meyer, 1857) (type)
P. hystricosus Schweitzer & Feldmann, 2009b
P. major (Hée, 1924)
P. rhathamingus Schweitzer & Feldmann, 2009b
P. schweigerti Schweitzer & Feldmann, 2009d
Nodoprosopidae Schweitzer & Feldmann, 2009a
Nodoprosopon†† Beurlen, 1928b (0, 0, 1)
N. ornatum (Von Meyer, 1860) (type)
Superfamily Dromioidea De Haan, 1833
Family **Dromiidae**† De Haan, 1833
Subfamily Dromiinae De Haan, 1833
Ameridromia†† Blow & Manning, 1996 (0, 0, 1)
A. hyneorum Blow & Manning, 1996 (type)
Cryptodromia† Stimpson, 1858a (22, 1, 0)
C. fallax† (Lamarck, 1818)
Dromia† Weber, 1795 (8, 0, 2)

D. anomala†† Rathbun, 1935a
D. neogenica†† Müller, 1978
Dromidia† Stimpson, 1858a (1, 0, 1)
D. bedetteae† Blow & Manning, 1996
Dromilites†† H. Milne Edwards, 1837 (0, 0, 12)
D. alpina Glaessner, 1929a
D. americana Rathbun, 1935a
D. bucklandi H. Milne Edwards, 1837 [in 1834-1840] (type)
D. corvini (Bittner, 1893)
D. fossata Müller & Collins, 1991
D. hilarionis (Bittner, 1883)
D. humerosus Quayle & Collins, 1981
D. lothi Förster & Mundlos, 1982
D. pastoris Vía, 1959
D. pentagonalis (Beurlen, 1928)
D. simplex Quayle & Collins, 1981
D. subglobosa Müller & Collins, 1991
D. vicensis Barnolas Cortinas, 1973
Epigodromia† McLay, 1993 (9, 1, 0)
E. areolata† (Ihle, 1913)
Kerepesia†† Müller, 1976 (0, 0, 1)
K. viai Müller, 1976 (type)
Kierionopsis†† Davidson, 1966 (0, 0, 1)
K. nodosa Davidson, 1966 (type)
Kromtitis†† Müller, 1984a (0, 0, 7)
K. koberi (Bachmayer & Tollmann, 1953) (type)
K. koberiformis Beschin, Busulini, De Angeli & Tessier, 2007
K. levigatus Beschin, Busulini, De Angeli & Tessier, 2007
K. pentagonalis Müller & Collins, 1991
K. spinulata Portell & Collins, 2004
K. subovatus Beschin, Busulini, De Angeli & Tessier, 2007
K. tetratuberculatus Beschin et al., 2002
Lucanthonisia†† Van Bakel et al., 2008 [imprint 2009] (0, 0, 2)

L. eotvoesi (Müller, 1976)
L. praemia Van Bakel et al., 2008 [imprint 2009] (type)
Noetlingia†† Beurlen, 1928b (0, 0, 3)
N. claudiopolitana Bittner, 1893 (type)
N. succini (Noetling, 1885a)
N. veronensis (Bittner, 1886)
Subfamily Hypoconchinae Guinot & Tavares, 2003
No fossil members
Subfamily Sphaerodromiinae Guinot & Tavares, 2003
No fossil members
Dynomenidae† Ortmann, 1892b (subfamilies not recognized for fossils at this time)
Acanthodiaulax†† Schweitzer, Feldmann et al., 2003 (0, 0, 1)
A. mclayi Schweitzer, Feldmann et al., 2003 (type)
Basinotopus†† McCoy, 1849 (0, 0, 2)
B. lamarckii (Desmarest, 1822) (type)
B. tricornis Collins & Jakobsen, 2003
Cyamocarcinus†† Bittner, 1883 (0, 0, 1)
C. angustifrons Bittner, 1883 (type) (= *C. budensis* Oppenheim, 1899)
Dromiopsis†† Reuss, 1858 [imprint 1857] (0, 0, 13)
D. americana Roberts, 1956
D. depressa Segerberg, 1900
D. elegans Reuss, 1858 [imprint 1857]
D. gigas Forir, 1887
D. granulata Collins & Rasmussen, 1992
D. kimberlyae Bishop, 1987
D. laevior Reuss, 1859
D. mosae Collins et al., 1995
D. paucigranosa Beschin, Busulini, De Angeli & Tessier, 2007
D. praelaevior Collins et al., 1995
D. pulchella Secretan, 1964
D. rugosus (Schlotheim, 1820) (type)
D. ubaghsi Forir, 1889
Dynomene† Desmarest, 1822 (5, 2, 5)
D. acutispinis†† (Rathbun, 1945)

D. emiliae†† Müller, 1978
D. hispida† Guérin-Méneville, 1832
D. lessinea†† Beschin et al., 2001
D. praedator† A. Milne-Edwards, 1878a
D. shinobui†† Karasawa, 1993
D. variabilis†† Portell & Collins, 2004
Dynomenopsis†† Secretan, 1972 (0, 0, 1)
D. branisai Secretan, 1972 (type)
Eotrachynotocarcinus†† Beschin, Busulini, De Angeli & Tessier, 2007 (0, 0, 1)
E. airaghii Beschin et al., 2007 (type)
Gemmellarocarcinus†† Checchia-Rispoli, 1905 (0, 0, 1)
G. loerentheyi Checchia-Rispoli, 1905 (type)
Graptocarcinus†† Roemer, 1887 (0, 0, 4)
G. bellonii Collins & Dieni, 1995
G. maastrichtensis Fraaye, 1996a
G. muiri Stenzel, 1944a
G. texanus Roemer, 1887 (type) (= *Cyphonotus integrimarginatus* Wright & Wright, 1950)
Maurimia†† Martins-Neto, 2001 (0, 0, 1)
M. sergipensis (Beurlen, 1965) (type)
Ovamene†† Müller & Collins, 1991 (0, 0, 1)
O. franciae Müller & Collins, 1991 (type)
Stephanometopon†† Bosquet, 1854 (0, 0, 1)
S. granulatum Bosquet, 1854 (type)
Trechmannius†† Collins & Donovan, 2006 (0, 0, 1)
T. circularis Collins & Donovan, 2006 (type)
Xandaros†† Bishop, 1988c (0, 0, 1)
X. sternbergi (Rathbun, 1926b) (type)
Diaulacidae†† Wright & Collins, 1972
Diaulax†† Bell, 1863 (0, 0, 7)
D. carteriana Bell, 1863 (type)
D. oweni (Bell, 1850)
D. italica Beschin, Busulini, De Angeli & Tessier, 2007
D. feliceps Wright & Collins, 1972
D. millerae Bishop, 1992a
D. roddai Bishop, 1983a
D. yokoii Collins et al., 1993

Superfamily Etyoidea†† Guinot & Tavares, 2001

Etyidae†† Guinot & Tavares, 2001

Caloxanthus†† A. Milne-Edwards, 1864 (= *Carpiliopsis*†† Fischer-Benzon, 1866; *Creticarcinus*†† Withers, 1928) (0, 0, 6)

C. americanus Rathbun, 1935a

C. kuypersi Fraaye, 1996a

C. formosus A. Milne-Edwards, 1864 (type)

C. ornata (Fischer-Benzon, 1866)

C. purleyensis (Withers, 1928)

C. simplex (Secretan, 1964)

Etyus†† Leach *in* Mantell, 1822 (0, 0, 1) (= *Reussia* McCoy, 1854)

E. martini Mantell, 1844 (type)

Etyxanthosia†† Fraaije, Van Bakel, Jagt & Artal, 2008 (0, 0, 4)

E. aspera (Rathbun, 1935a)

E. fossa (Wright & Collins, 1972) (type)

E. pawpawensis (Schweitzer, Salva & Feldmann, 1999)

E. reidi (Schweitzer, Salva & Feldmann, 1999)

Feldmannia†† Guinot & Tavares, 2001 (0, 0, 2)

F. jacksoni (Wright & Collins, 1972)

F. wintoni (Rathbun, 1935a) (type)

Guinotosia†† Beshin, Busulini, De Angeli & Tessier, 2007 (0, 0, 1)

G. tertiaria Beschin et al., 2007 (type)

Sharnia†† Collins & Saward, 2006 (0, 0, 1)

S. burnhamensis Collins & Saward, 2006 (type)

Xanthosia†† Bell, 1863 (0, 0, 5)

X. buchii (Reuss, 1845)

X. gibbosa Bell, 1863 (type)

X. granulosa (McCoy, 1854)

X. sakoi Karasawa, Ohara & Kato, 2008

X. similis (Bell, 1863)

Superfamily Homoloidea De Haan, 1839

Family **Gastrodoridae**†† Van Bakel et al., 2008

Gastrodorus†† Von Meyer, 1864 (0, 0, 2) (= *Eogastrodorus* Van Bakel et al., 2008)

G. granulatus Förster, 1985
G. neuhausense Von Meyer, 1864 (type)
Family **Homolidae**† De Haan, 1839
Dagnaudus† Guinot & Richer de Forges, 1995 (1, 0, 1)
D. pritchardi†† (Jenkins, 1977)
Doerflesia†† Feldmann & Schweitzer, 2009 (0, 0, 1)
D. ornata Feldmann & Schweitzer, 2009 (type)
Homola† Leach, 1816 [imprint 1815] (11, 0, 2)
H. bajaensis†† Schweitzer, Feldmann, Gonzáles-Barba & Ćosović, 2006
H. vancouverensis†† Schweitzer, 2001
Homoliformis†† Collins, Schulz & Jakobsen, 2005 (0, 0, 1)
H. vagus Collins et al., 2005 (type)
Homolopsis†† Bell, 1863 (0, 0, 10)
H. chilensis Förster & Stinnesbeck, 1987
H. depressa Carter, 1898
H. edwardsii Bell, 1863 (type)
H. gibbosa (Schlüter, 1879)
H. glabra Wright & Collins, 1972
H. hachiyai Takeda & Fujiyama, 1983
H. spinosa Van Straelen, 1936a
H. spinulosa Glaessner, 1980
H. tuberculata Van Straelen, 1936a
H. williamsi Bishop, 1992b
Hoplitocarcinus†† Beurlen, 1928b (0, 0, 1)
H. johannesboehmi Beurlen, 1928b (type)
Latheticocarcinus†† Bishop, 1988b (0, 0, 15)
L. adelphinus (Collins & Rasmussen, 1992)
L. affinus (Jakobsen & Collins, 1997)
L. atlanticus (Roberts, 1962)
L. brevis (Collins, Kanie & Karasawa, 1993)
L. brightoni (Wright & Collins, 1972)
L. centurialis (Bishop, 1992b)
L. declinatus (Collins, Fraaye & Jagt, 1995)
L. dispar (Roberts, 1962)
L. ludvigseni Schweitzer, Nyborg, Feldmann & Ross, 2004

L. pikeae (Bishop & Brannen, 1992)
L. punctatus (Rathbun, 1917a)
L. schlueteri (Beurlen, 1928b)
L. shapiroi Bishop, 1988b (type)
L. spinigus (Jakobsen & Collins, 1997)
L. transiens (Segerberg, 1900)
Lignihomola†† Collins, 1997 (0, 0, 1)
L. etheridgei (Woodward, 1892) (type)
Londinimola†† Collins & Saward, 2006 (0, 0, 1)
L. williamsi Collins & Saward, 2006 (type)
Paromola† Wood-Mason *in* Wood-Mason & Alcock, 1891 (6, 0, 1)
P. vetula†† Crawford, 2008
Paromolopsis† Wood-Mason *in* Wood-Mason & Alcock, 1891 (1, 0, 1)
P. piersoni†† Schweitzer, Nyborg, Feldmann & Ross, 2004
Prohomola†† Karasawa, 1992a (0, 0, 2)
P. japonica (Yokoyama, 1911) (type)
P. katunai Blow & Manning, 1996
Zygastrocarcinus†† Bishop, 1983b (0, 0, 4)
Z. cardsmithi Bishop, 1986b
Z. griesi Bishop, 1983b (type)
Z. mendryki (Bishop, 1982)
Z. waagei Feldmann, Schweitzer & Green, 2008

Latreilliidae Stimpson, 1858a
Questionably *Heeia*†† Wright & Collins, 1972 (0, 0, 1)
H. villersensis (Hée, 1924) (= *Calappa cranium* Wright & Wright, 1945) (type)

Poupiniidae† Guinot, 1991
Rhinopoupinia†† Feldmann, Tshudy & Thomson, 1993 (0, 0, 1)
R. bicornis Feldmann et al., 1993 (type)

Tithonohomolidae†† Feldmann & Schweitzer, 2009
Tenuihomola†† Feldmann & Schweitzer, 2009 (0, 0, 2)
T. longa (Moericke, 1889) (type)
T. ortwini Feldmann & Schweitzer, 2009
Tithonohomola†† Glaessner, 1933a (0, 0, 2)

T. armata (Blaschke, 1911) (type)
T. tuberculata Feldmann & Schweitzer, 2009
Section Eubrachyura de Saint Laurent, 1980
Family **Marocarcinidae**†† Guinot, De Angeli & Garassino, 2008
Marocarcinus†† Guinot, De Angeli & Garassino, 2008 (0, 0, 1)
M. pasinii Guinot, De Angeli & Garassino, 2008 (type)
Subsection Raninoida De Haan, 1839
Superfamily Raninoidea De Haan, 1839
Family **Camarocarcinidae**†† Feldmann, Li & Schweitzer, 2007
Camarocarcinus†† Holland & Cvancara, 1958 (0, 0, 3)
C. arnesoni Holland & Cvancara, 1958 (type)
C. obtusus Jakobsen & Collins, 1979
C. quinquetuberculatus Collins & Rasmussen, 1992
Cretacocarcinus†† Feldmann, Li & Schweitzer, 2007 (0, 0, 1)
C. smithi Feldmann et al., 2007 (type)
Family **Cenomanocarcinidae** Guinot, Vega & Van Bakel, 2008
Campylostoma†† Bell, 1858 (0, 0, 1)
C. matutiforme Bell, 1858 (type)
Cenomanocarcinus†† Van Straelen, 1936a (0, 0, 9) (=*Sagittiformosus*†† Bishop, 1988b)**
C. armatus (Rathbun, 1935a)
C. beardi Schweitzer, Feldmann et al., 2003
C. hierosolymitanus Avnimelech, 1961
C. inflatus Van Straelen, 1936a (type)
C. multituberculatus (Joleaud & Hsu, 1935)
C. oklahomensis (Rathbun, 1935a)
C. renfroae (Stenzel, 1945)
C. vanstraeleni Stenzel, 1945
C. siouxensis Feldmann et al., 1976
Family **Raninidae**† De Haan, 1839
Subfamily uncertain
Araripecarcinus†† Martins-Neto, 1987 (0, 0, 1): This is a raninid sternum (see Karasawa, Schweitzer & Feldmann, 2008)

A. ferreirai Martins-Neto, 1987 (type)

Subfamily Cyrtorhininae† Guinot, 1993

Cyrtorhina† Monod, 1956 (2, 0, 4)

C. fusseli†† Blow & Manning, 1996

C. globosa†† Beschin, Busulini, De Angeli & Tessier, 1988

C. oblonga†† Beschin, Busulini, De Angeli & Tessier, 1988

C. ripacurtae†† Artal & Castillo, 2005a

Subfamily Lyreidinae† Guinot, 1993

Hemioon†† Bell, 1863 (0, 0, 7)

H. bidentata (Rathbun, 1935a)

H. circumviator Wright & Collins, 1972

H. cunningtoni Bell, 1863 (type)

H. elongatum (A. Milne-Edwards, 1862b)

H. eysunesensis Collins & Rasmussen, 1992

H. novozelandicum Glaessner, 1980

H. yanini Ilyin & Alekseev, 1998

Lyreidina†† Fraaye & Van Bakel, 1998 (0, 0, 1)

L. pyriformis Fraaye & Van Bakel, 1998 (type)

Lyreidus† De Haan, 1841 (=*Lysirude* Goeke, 1985) (6, 0, 10)

L. antarcticus†† Feldmann & Zinsmeister, 1984b

L. bennetti†† Feldmann & Maxwell, 1990

L. elegans†† Glaessner, 1960

L. fastigatus†† Rathbun, 1919 (only pereiopod)

L. hookeri†† Feldmann, 1992

L. hungaricus†† Beurlen, 1939

L. lebuensis†† Feldmann, 1992

L. paronae†† Crema, 1895

L. teshimai†† (Fujiyama & Takeda, 1980)

L. waitakiensis†† Glaessner, 1980

Rogueus†† Berglund & Feldmann, 1989 (0, 0, 2)

R. orri Berglund & Feldmann, 1989 (type)

R. robustus Collins & Jakobsen, 1995

Macroacaena†† Tucker, 1998 (=*Carinaranina* Tucker, 1998) (0, 0, 12)

M. alseana (Rathbun, 1932a)

M. bispinulata (Collins & Rasmussen, 1992)
M. chica Schweitzer, Feldmann et al., 2003
M. franconia Schweigert et al., 2004
M. fudoujii (Karasawa, 2000a)
M. leucosiae (Rathbun, 1932a)
M. marionae (Tucker, 1998)
M. naselensis (Rathbun, 1926b)
M. rosenkrantzi (Collins & Rasmussen, 1992)
M. schencki (Rathbun, 1932a)
M. succedana (Collins & Rasmussen, 1992) type
M. willapensis (Rathbun, 1926b)
Tribolocephalus†† Ristori, 1886 (0, 0, 1)
T. laevis Ristori, 1886 (type)
Subfamily Notopodinae† Serène & Umali, 1972
Erroranilia†† Boyko, 2004 (0, 0, 1)
E. taipeiensis (Hu & Tao, 1996) (type)
Eumorphocorystes†† Van Binkhorst, 1857 (=*Pseudoraninella* Lőrenthey *in* Lőrenthey & Beurlen, 1929) (0, 0, 4)
E. muelleri (Van Binkhorst, 1857)
E. sculptus Van Binkhorst, 1857 (type)
E. toehoepae (Van Straelen, 1924 [imprint 1923])
E. vahldieki (Förster & Mundlos, 1982)
Lianira†† Beschin, Busulini, De Angeli, Tessier & Ungaro, 1991 (0, 0, 3)
L. beschini Beschin et al., 1991 (type)
L. convexa Beschin et al., 1991
L. isidoroi Beschin, De Angeli & Checchi, 2007
Lovarina†† Beschin, Busulini, De Angeli, Tessier & Ungaro, 1991 (0, 0, 1)
L. cristata Beschin et al., 1991 (type)
Notopus† De Haan, 1841 (0, 1, 2)
N. beyrichi†† Bittner, 1875
N. dorsipes† (Linnaeus, 1758) [*Dorippe dorsipes* [*sensu*] Glaessner, 1929b]
N. minutus†† Vega, Cosma et al., 2001
Ponotus†† Karasawa & Ohara, 2009 (0, 0, 1)
P. shirahamensis Karasawa & Ohara, 2009 (type)

Ranidina†† Bittner, 1893 (0, 0, 1)
R. rosaliae Bittner, 1893 (type)
Raniliformis†† Jagt, Collins & Fraaye, 1993 (0, 0, 7)
R. baltica (Segerberg, 1900) (type)
R. chevrona Fraaye & Van Bakel, 1998
R. eocaenica (Beschin, Busulini, De Angeli & Tessier, 1988)
R. occlusa Collins, Fraaye & Jagt, 1995
R. ornata Beschin, De Angeli & Checchi, 2007
R. prebaltica Fraaye & Van Bakel, 1998
R. rugosa Beschin, De Angeli & Checchi, 2007
Ranilia† H. Milne Edwards, 1837 [in 1834-1840] (=*Notopella* Lőrenthey *in* Lőrenthey & Beurlen, 1929; =*Raninops* A. Milne-Edwards, 1880a) (6, 1, 2)
R. constricta† (A. Milne-Edwards, 1880a)
R. punctulata†† Beschin, Busulini, De Angeli & Tessier, 1988
R. vareolata†† (Lőrenthey *in* Lőrenthey & Beurlen, 1929)
Umalia† Guinot, 1993 (6, 1, 1)
U. guinotae†† De Angeli & Beschin, 2007
U. misakiensis† (Sakai, 1937)
Subfamily Ranininae† De Haan, 1839
Lophoranina†† Fabiani, 1910 (0, 0, 24)
L. aculeata (A. Milne-Edwards, 1881b)
L. aldrovandii (Ranzani, 1820)
L. bakerti (A. Milne-Edwards, 1872)
L. barroisi (Brocchi, 1877)
L. bishopi Squires & Demetrion, 1992
L. bittneri (Lőrenthey, 1902)
L. cristaspina Vega, Cosma et al., 2001
L. georgiana (Rathbun, 1935a)
L. kemmelingi Van Straelen, 1924 [imprint 1923]
L. laevifrons (Bittner, 1875)
L. levantina Lewy, 1977
L. marestiana (König, 1825) (type)
L. maxima Beschin et al., 2004
L. persica Withers, 1932b

L. porifera (Woodward, 1866c)
L. precocious Feldmann et al., 1996
L. quinquespinosa (Rathbun, 1945)
L. raynorae Blow & Manning, 1996
L. reussi (Woodward, 1866c)
L. rossi Blow & Manning, 1996
L. soembaensis Van Straelen, 1938
L. straeleni Vía, 1959
L. tchihatcheffi (A. Milne-Edwards, 1866)
L. toyosimai Yabe & Sugiyama, 1935
Ranina† Lamarck, 1801 (0, 1, 17)
R. americana†† Withers, 1924b
R. berglundi†† Squires & Demetrion, 1992
R. bouilleana†† A. Milne-Edwards, 1872
R. brevispina†† Lőrenthey, 1898b
R. burleighensis†† Holland & Cvancara, 1958
R. cuspidata†† Guppy, 1909
R. elegans†† Rathbun, 1945
R. granulosa†† A. Milne-Edwards, 1872
R. haszlinskyi†† Reuss, 1859
R. hirsuta†† Schafhäutl, 1863
R. lamiensis†† Rathbun, 1945
R. molengraaffi†† Van Straelen, 1924 [imprint 1923]
R. oblonga†† Münster, 1840
R. palmea†† E. Sismonda, 1846
R. propinqua†† Ristori, 1891
R. ranina† (Linnaeus, 1758) (type)
R. speciosa†† (Münster, 1840)
R. tejoniana†† Rathbun, 1926b
Raninella†† A. Milne-Edwards, 1862b (0, 0, 15)
R. armata Rathbun, 1935a
R. atava Carter, 1898
R. carlilensis Feldmann & Maxey, 1980
R. eocenica Rathbun, 1935a
R. griesbachi (Noetling, 1897)
R. libyca Van Straelen, 1935 [imprint 1934]
R. manningi Bishop & Williams, 2000
R. mucronata Rathbun, 1935a

R. oaheensis Bishop, 1978
R. quadrispinosus (Collins, Fraaye & Jagt, 1995)
R. starkvillensis Rathbun, 1935a
R. testacea Rathbun, 1926a
R. trechmanni Withers, 1927
R. tridens Roberts, 1962
R. trigeri A. Milne-Edwards, 1862b (type)
Subfamily Raninoidinae Lőrenthey *in* Lőrenthey & Beurlen, 1929
Bicornisranina†† Nyborg & Fam, 2008 (0, 0, 1)
B. bocki Nyborg & Fam, 2008 (type)
Cenocorystes†† Collins & Breton, 2009 (0, 0, 1)
C. fournieri Collins & Breton, 2009 (type)
Cristafrons†† Feldmann, Tshudy & Thomson, 1993 (0, 0, 1)
C. praescientis Feldmann et al., 1993 (type)
Notopoides† Henderson, 1888 (1, 0, 3)
N. exiguus†† Beschin, Busulini, De Angeli & Tessier, 1988
N. pflugervillensis†† Beikirch & Feldmann, 1980
N. verbeeki†† Böhm, 1922
Notosceles† Bourne, 1922 (4, 1, 0)
N. serratifrons† (Henderson, 1893)
Quasilaeviranina†† Tucker, 1998 (0, 0, 6)
Q. arzignanensis (Beschin, Busulini, De Angeli & Tessier, 1988)
Q. keyesi (Feldmann & Maxwell, 1990)
Q. ombonii (Fabiani, 1910)
Q. ovalis (Rathbun, 1935a)
Q. pororariensis (Glaessner, 1980)
Q. simplicissima (Bittner, 1883)
Raninoides† H. Milne Edwards, 1837 [in 1834-1840] (=*Laeviranina* Lőrenthey *in* Lőrenthey & Beurlen, 1929; =*Pseudorogueus*†† Fraaye, 1995; =*Raninellopsis* Böhm, 1922) (9, 2, 32)
R. acanthocolus†† Schweitzer, Feldmann, Gonzáles-Barba & Ćosović, 2006
R. araucana†† (Philippi, 1887)

R. asper†† Rathbun, 1926b
R. barnardi† Sakai, 1974
R. benedicti† Rathbun, 1935b
R. borealis†† (Collins & Rasmussen, 1992)
R. budapestiniensis†† (Lőrenthey, 1897)
R. dickersoni†† Rathbun, 1926b
R. eugenensis†† Rathbun, 1926b
R. fabianii†† (Lőrenthey *in* Lőrenthey & Beurlen, 1929)
R. fulgidus†† Rathbun, 1926b
R. fulungensis†† Hu & Tao, 1999
R. glabra†† (Woodward, 1871)
R. goedertorum†† (Tucker, 1998)
R. gottschei†† Böhm, 1927
R. hollandica†† (Collins et al., 1997)
R. javanus†† (Böhm, 1922)
R. madurensis†† Beets, 1950
R. mexicanus†† Rathbun, 1930b
R. morrisi†† Collins et al., 2003
R. nodai†† Karasawa, 1992a
R. notopoides†† (Bittner, 1883)
R. oregonensis†† Rathbun, 1926b
R. perarmata†† (Glaessner, 1960)
R. proracanthus†† Schweitzer, Feldmann, Gonzáles-Barba & Ćosović, 2006
R. pulchra†† (Beschin, Busulini, De Angeli & Tessier, 1988)
R. rangiferus†† (Fraaye, 1995)
R. rathbunae†† Van Straelen, 1933b
R. sinuosus†† (Collins & Morris, 1978)
R. slaki†† Squires, 2001
R. treldenaesensis†† Collins & Jakobsen, 2003
R. vaderensis†† Rathbun, 1926b (= *Laeviranina lewisana* Rathbun, 1926b)
R. washburnei†† Rathbun, 1926b

Subfamily Palaeocorystinae Lőrenthey *in* Lőrenthey & Beurlen, 1929
Cretacoranina†† Mertin, 1941 (0, 0, 7)
C. dichrous (Stenzel, 1945)

C. laevis (Schlüter *in* Schlüter & Von der Marck, 1868)
C. ornatus Wright & Collins, 1972
C. punctata (Rathbun, 1935a)
C. paututensis Collins & Rasmussen, 1992
C. schloenbachi (Schlüter, 1879) (type)
C. syriacus Withers, 1928
Eucorystes†† Bell, 1863 (0, 0, 8)
E. broderipi (Mantell, 1844)
E. carteri McCoy, 1854 (type)
E. eichhorni Bishop, 1983b
E. harveyi (Woodward, 1896)
E. intermedius Nagao, 1931
E. mangyshlakensis Ilyin & Pistshikova *in* Ilyin, 2005
E. platys Schweitzer & Feldmann, 2001a
E. oxtedensis Wright & Collins, 1972
Heus†† Bishop & Williams, 2000 (0, 0, 1)
H. foersteri Bishop & Williams, 2000 (type)
Notopocorystes†† McCoy, 1849 (0, 0, 14)
N. australis Secretan, 1964
N. beyrichi Bittner, 1875
N. bituberculatus Secretan, 1964
N. callianassarum (Fritsch & Kafka, 1887)
N. denisae Secretan, 1964
N. exiguus Glaessner, 1980
N. fritschi Glaessner, 1929b
N. japonicus (Jimbô, 1894)
N. normani (Bell, 1863)
N. punctatus Rathbun, 1935a
N. parvus Rathbun, 1935a
N. ripleyensis Rathbun, 1935a
N. stokesii (Mantell, 1844) (type)
N. xizangensis Wang, 1981
Subfamily indeterminate
Sabahranina†† Collins et al., 2003 (0, 0, 1)
S. trushidupensis Collins et al., 2003 (type)
"*Symnista*" *bidentata*†† Rathbun, 1935a
Symethidae† Goeke, 1981
Symethis† Weber, 1795 (3, 0, 1)

S. johnsoni†† Rathbun, 1935a

Superfamily Cyclodorippoidea Ortmann, 1892b

Family **Cyclodorippidae**† Ortmann, 1892b

Subfamily Cyclodorippinae Ortmann, 1892b

Tymolus† Stimpson, 1858b (9, 0, 4)

T. alaskensis†† Conkle et al., 2006

T. ingens†† Takeda & Tomida, 1984

T. itoigawai†† Takeda & Tomida, 1984

T. kamadai†† Imaizumi, 1952b

Subfamily Xeinostomatinae Tavares, 1992

Xeinostoma† Stebbing, 1920 (4, 0, 1?)

X. ? *antiqua*†† Schweitzer, 2001

Cymonomidae† Bouvier, 1897

Cymonomus† A. Milne-Edwards, 1880a (24, 0, 1)

C. primitivus†† Müller & Collins, 1991

Phyllotymolinidae Tavares, 1998

No fossil representatives

Torynommatidae†† Glaessner, 1980

Binkhorstia†† Noetling, 1881 (0, 0, 2)

B. euglypha Collins et al., 1995

B. ubaghsii (Van Binkhorst, 1857) (type)

Dioratiopus†† Woods, 1953 (0, 0, 12) (= *Glaessnerella* Wright & Collins, 1975; *Glaessneria* Wright & Collins, 1972)

D. angusta (Wright & Collins, 1972)

D. balli (Wright & Collins, 1972)

D. dawsonensis (Bishop, 1973)

D. decipiens (Wright & Collins, 1972)

D. depressa (Carter, 1898)

D. hearttailensis Bishop, 1985a

D. kennedyi (Wright & Collins, 1972)

D. latteri (Wright & Collins, 1972)

D. primitivus Fraaye & Collins, 1996

D. salebrosus Woods, 1953 (type)

D. scotti Bishop, 1983a

D. spinosa (Van Straelen, 1936a)

Torynomma†† Woods, 1953 (0, 0, 6)

T. australis Feldmann, Tshudy & Thomson, 1993

T. densus Bishop, 1983a

T. dentatum Glaessner, 1980

T. flemingi Glaessner, 1980

T. planata Feldmann, 1993

T. quadrata Woods, 1953 (type)

Withersella†† Wright & Collins, 1972 (0, 0, 1)

W. crepitans Wright & Collins, 1972 (type)

Subsection Heterotremata Guinot, 1977

Superfamily Componocancroidea†† Feldmann, Schweitzer & Green, 2008

Family **Componocancridae**†† Feldmann et al., 2008

Componocancer†† Feldmann et al., 2008 (0, 0, 1)

C. roberti Feldmann et al., 2008 (type)

Superfamily Dorippoidea MacLeay, 1838

Family **Dorippidae**† MacLeay, 1838

Archaeocypoda†† Secretan, 1975 (0, 0, 1)

A. veronensis Secretan, 1975 (type)

Bartethusa†† Quayle & Collins, 1981 (0, 0, 1)

B. hepatica Quayle & Collins, 1981 (type)

Dorippe† Weber, 1795 (7, 0, 3)

D. frankhauseri†† Studer, 1898

D. margaretha†† Lőrenthey *in* Lőrenthey & Beurlen, 1929

D. ornatissima†† Müller, 2006

Eodorippe†† Glaessner, 1980 (0, 0, 1)

E. spedeni Glaessner, 1980 (type)

Heikeopsis† Ng, Guinot & Davie, 2008 (2, 0, 1) (replacement name for *Heikea* Holthuis & Manning, 1990)

H. tuberculata†† (Morris & Collins, 1991)

Hillius†† Bishop, 1983a (0, 0, 1)

H. youngi Bishop, 1983a (type)

Medorippe† Manning & Holthuis, 1981 (0, 1, 2)

M. ampla†† Garassino et al., 2004

M. lanata† (Linnaeus, 1767)

M. tanabei†† Karasawa, 2000b

Neodorippe† Serène & Romimohtarto, 1969 (2, 0, 1?)

N. ? carpathica†† (Förster, 1979)

Nobilum† Serène & Rohmimohtarto, 1969 (1, 0, 1)

N. wenchii†† Hu & Tao, 1996
Paradorippe† Serène & Romimohtarto, 1969 (3, 1, 0)
P. granulata† (De Haan, 1839)
Sodakus†† Bishop, 1978 (0, 0, 2)
S. mexicanus Vega, Feldmann & Villalobos-Hiriart, 1995
S. tatankayotankaensis Bishop, 1978 (type)
Telamonocarcinus†† Larghi, 2004 (0, 0, 2)
T. binodosus (Collins, Kanie & Karasawa, 1993)
T. gambalatus Larghi, 2004 (type)
Tepexicarcinus†† Feldmann, Vega, Applegate & Bishop, 1998 (0, 0, 1)
T. tlayuaensis Feldmann et al., 1998 (type)
Titanodorippe†† Blow & Manning, 1996 (0, 0, 1)
T. eocenica Blow & Manning, 1996 (type)
Ethusidae† Guinot, 1977
Ethusa† Roux, 1830 (42, 1, 5)
E. berica†† De Angeli & Beschin, 2008
E. chibai†† Karasawa, 1993
E. evae†† Müller & Collins, 1991
E. mascarone† (Herbst, 1785)
E. octospinosa†† Müller, 2006
Orithyiidae Dana, 1852a
No fossil representatives
Necrocarcinidae†† Förster, 1968
Corazzatocarcinus†† Larghi, 2004 (0, 0, 1)
C. hadjoulae (Roger, 1946) (type)
Glyptodynomene†† Van Straelen, 1944 (0, 0, 2)
G. alsasuensis Van Straelen, 1944 (type)
G. inornata Collins et al., 1995
Necrocarcinus†† Bell, 1863 (0, 0, 21)
N. angelicus Fraaije, 2002b
N. avicularis Fritsch & Kafka, 1887
N. bedrakensis Levitski, 1974
N. bispinosus Segerberg, 1900
N. carinatus Feldmann, Tshudy & Thomson, 1993
N. davisi Bishop, 1985a
N. insignis Segerberg, 1900

N. isericus (Fritsch & Kafka, 1887)
N. labeschei (Eudes-Deslongchamps, 1835) (type)
N. olsonorum Bishop & Williams, 1991
N. olssoni (Rathbun, 1937)
N. ornatissimus Forir, 1887
N. perlatus Fritsch & Kafka, 1887
N. pierrensis (Rathbun, 1917a)
N. rathbunae Roberts, 1962
N. senoniensis Schlüter *in* Schlüter & Von der Marck, 1868
N. tauricus Ilyin & Alekseev, 1998
N. texensis Rathbun, 1935a
N. undecimtuberculatus Takeda & Fujiyama, 1983
N. woodwardi Bell, 1863
N. wrighti Feldmann, Tshudy & Thomson, 1993
Paranecrocarcinus†† Van Straelen, 1936a (0, 0, 10)
P. digitatus Wright & Collins, 1972
P. foersteri Wright & Collins, 1972
P. graysonensis (Rathbun, 1935a)
P. hexagonalis Van Straelen, 1936a (type)
P. kennedyi Wright, 1997
P. libanoticus Förster, 1968
P. moseleyi (Stenzel, 1945)
P. mozambiquensis Förster, 1970a
P. ovalis (Stenzel, 1945)
P. vanbirgeleni Fraaije, 2002b
Polycnemidium†† Reuss, 1855 (0, 0, 1)
P. pustulosum (Reuss, 1845) (type)
Pseudonecrocarcinus†† Förster, 1968 (0, 0, 5)
P. biscissus Wright & Collins, 1972
P. gamma (Roberts, 1962)
P. quadriscissus (Noetling, 1881) (type)
P. scotti (Stenzel, 1945)
P. stenzeli Bishop, 1983a
Shazella†† Collins & Williams, 2004 (0, 0, 1)
S. abbotsensis Collins & Williams, 2004 (type)
Orithopsidae†† Schweitzer, Feldmann et al., 2003
Cherpiocarcinus†† Marangon & De Angeli, 1997 (0, 0, 1)

C. rostratus Marangon & De Angeli, 1997 (type)
Goniochele†† Bell, 1858 (0, 0, 3)
G. angulata Bell, 1858 (type)
G. armata Rathbun, 1918
G. madseni Collins & Jakobsen, 2003
Marycarcinus†† Schweitzer, Feldmann et al., 2003 (0, 0, 1)
M. hannae (Rathbun, 1926b) (type)
Orithopsis†† Carter, 1872 (0, 0, 2)
O. bonneyi Carter, 1872 (type)
O. tricarinatus (Woodward, 1868a)
Paradoxicarcinus†† Schweitzer, Feldmann et al., 2003 (0, 0, 1)
P. nimonoides Schweitzer et al., 2003 (type)
Silvacarcinus†† Collins & Smith, 1992 (0, 0, 1)
S. laurae Collins & Smith, 1992 (type)
Superfamily Calappoidea De Haan, 1833
Family uncertain
Cristella†† Collins & Rasmussen, 1992 (0, 0, 1)
C. hastata Collins & Rasmussen, 1992 (type)
Family **Calappidae**† De Haan, 1833
Acanthocarpus† Stimpson, 1870 (5, 0, 1)
A. obscurus†† (Rathbun, 1918)
Bittnerilia†† De Angeli & Garassino, 2003 (0, 0, 2)
B. dentata Beschin et al., 2005
B. eocenica (Bittner, 1883) (type)
Calappa† Weber, 1795 (34, 7, 18)
C. chungii†† Hu & Tao, 1985
C. costaricana†† Rathbun, 1918
C. earlei†† Withers, 1924c
C. flammea† (Herbst, 1794)
C. cf. *C. gallus*† (Herbst, 1803)
C. granulata† (Linnaeus, 1758) (type)
C. hepatica† (Linnaeus, 1758)
C. heberti†† Brocchi, 1883
C. lanensis†† Rathbun, 1926b
C. laraensis†† Van Straelen, 1933b
C. lophos† (Herbst, 1795)

C. madoerensis†† Van Straelen, 1938

C. oboui†† Hu & Tao, 1996

C. pavimenta†† Schweitzer, Iturralde-Vinent, Hetler & Velez-Juarbe, 2006

C. praelata†† Lőrenthey *in* Lőrenthey & Beurlen, 1929

C. protopustulosa†† Noetling, 1901

C. pustulosa† Alcock, 1896

C. restricta†† A. Milne-Edwards, 1873a

C. robertsi†† Ross et al., 1964

C. sahelensis†† Van Straelen, 1936d

C. sangiranensis†† Van Straelen, 1938

C. sexaspinosa†† Morris & Collins, 1991

C. springeri† Rathbun, 1931b

C. zinsmeisteri†† Feldmann & Wilson, 1988

C. zurcheri†† Bouvier, 1899

Calappella†† Rathbun, 1918 (0, 0, 1)

C. quadrispina Rathbun, 1918 (type)

Calappilia†† A. Milne-Edwards *in* de Bouillé, 1873 (0, 0, 23)

C. bohmi Glaessner, 1929b

C. bonairensis Van Straelen, 1933c

C. borneoensis Van Straelen, 1924 [imprint 1923]

C. brooksi Ross & Scolaro, 1964

C. ? chilensis Feldmann et al., 2005

C. circularis (Beurlen, 1958a)

C. dacica Bittner, 1893

C. diglypta Stenzel, 1934

C. gemmata Beschin et al., 1994

C. hondoensis Rathbun, 1930b

C. incisa Bittner, 1886

C. mainii Allasinaz, 1987

C. matzkei (Bachmayer, 1962)

C. maxwelli Feldmann, 1993

C. ocalanus (Ross et al., 1964)

C. perlata Noetling, 1885a

C. scopuli Quayle & Collins, 1981

C. sexdentata A. Milne-Edwards *in* Bouillé, 1876

C. sitzi Blow & Manning, 1996

C. subovata Beschin et al., 2002
C. tridentata (Beurlen, 1939)
C. verrucosa A. Milne-Edwards, 1873 (type)
C. vicetina Fabiani, 1910
Cryptosoma† Brullé *in* Barker-Webb & Berthelot, 1840 (3, 1, 0)
C. bairdii† (Stimpson, 1862b)
Cycloes† De Haan, 1837 (1, 1, 0)
C. granulosa† De Haan, 1837 (type)
Mursia† Leach *in* Desmarest, 1822 (23, 2, 12)
M. armata† De Haan, 1837
M. aspina†† Schweitzer & Feldmann, 2000b
M. australiensis† Campbell, 1971
M. bekenuensis†† Collins et al., 2003
M. circularis†† (Karasawa, 1989)
M. creutzbergi†† Collins & Donovan, 2004
M. granulosa†† Collins & Donovan, 2002
M. lienharti†† (Bachmayer, 1962)
M. macdonaldi†† Rathbun, 1918
M. marcusana†† Rathbun, 1926b
M. minuta†† Karasawa, 1993
M. obscura†† Rathbun, 1918
M. takahashii†† Imaizumi, 1952c
M. yaquinensis†† Rathbun, 1926b
Mursilata†† Hu & Tao, 1996 (0, 0, 3)
M. chuangi Hu & Tao, 1996
M. kilmeri (Hu, 1981) (type)
M. tzengi Hu & Tao, 1996
Mursilia†† Rathbun, 1918 (0, 0, 1)
M. ecristata Rathbun, 1918 (type)
Mursiopsis†† Ristori, 1889a (0, 0, 1)
M. pustulosus Ristori, 1889a (type)
Paracyclois† Miers, 1886 (2, 0, 1)
P. grandispinis†† Etheridge & McCullock, 1916
Stenodromia†† A. Milne-Edwards *in* Bouillé, 1873 (0, 0, 2)
S. calasanctii Vía, 1959
S. gibbosa A. Milne-Edwards *in* Bouillé, 1873 (type)

Tutus†† Collins *in* Collins et al., 2009 (0, 0, 1)
T. granulosus Collins *in* Collins et al., 2009 (type)

Family **Aethridae**† Dana, 1851b

Drachiella† Guinot *in* Serène & Soh, 1976 (6, 0, 1)
D. guinotae†† Morris & Collins, 1991

Eriosachila†† Blow & Manning, 1996 (0, 0, 8)
E. bajaensis Schweitzer, Feldmann, Gonzáles-Barba & Vega, 2002
E. bartholomaeensis (Rathbun, 1919)
E. cuvillieri (Van Straelen, 1929 [imprint 1930])
E. orri Schweitzer & Feldmann, 2000b
E. petiti Blow & Manning, 1996 (type)
E. rathbunae Maury, 1930
E. rossi Schweitzer & Feldmann, 2000b
E. terryi (Rathbun, 1937)

Hepatella† Smith, 1869a (2, 0, 1)
H. amazonica†† Beurlen, 1958a

Hepatiscus†† Bittner, 1875 (0, 0, 9)
H. distefanoi Checchia-Rispoli, 1905
H. laevis Lőrenthey *in* Lőrenthey & Beurlen, 1929
H. minimus Beschin et al., 1994
H. neumayri Bittner, 1875 (type)
H. poverelli Vía, 1959
H. pulchellus Bittner, 1875
H. schweinfurthi Noetling, 1885b
H. sheranii Collins & Morris, 1978
H. subpulchellus Böhm, 1922

Hepatus† Latreille, 1802 (5, 2, 7)
H. biformis†† Collins & Todd *in* Todd & Collins, 2005
H. bottomsi†† Blow, 2003
H. chiliensis† H. Milne Edwards, 1837 [in 1834-1840]
H. guraboensis†† Collins *in* Collins et al., 2009
H. lineatus† Rathbun, 1898a
H. lineatinus†† Collins & Todd *in* Todd & Collins, 2005
H. nodosus†† Collins & Morris, 1976
H. praecox†† Collins, Donovan & Dixon, 1996
H. spinimarginatus†† Feldmann et al., 2005

Mainhepatiscus†† De Angeli & Beschin, 1999 (0, 0, 1)

M. zannatoi De Angeli & Beschin, 1999 (type)
Matutites†† Blow & Manning, 1996 (0, 0, 3)
M. americanus (Rathbun, 1935a)
M. anthonyae Blow & Manning, 1996 (type)
M. miltonorum Feldmann, Bice et al., 1998
Osachila† Stimpson, 1871a (11, 0, 3)
O. berica†† De Angeli & Beschin, 1999
O. tiechi†† Studer, 1898
O. valdesensis†† Casadío et al., 2005
Prehepatus†† Rathbun, 1935a (0, 0, 8)
P. cretaceous Rathbun, 1935a (type)
P. dilksi Roberts, 1962
P. harrisi Bishop, 1985b
P. hodgesi Bishop, 1983a
P. lenolensis (Rathbun, 1935c)
P. mexicanus Schweitzer, Feldmann, González-Barba & V. Ćosović, 2006
P. pawpawensis Rathbun, 1935a
P. werneri Fraaye & Collins, 1987
Priabonella†† Beschin, De Angeli, Checchi & Mietto, 2006 (0, 0, 1)
P. violatii Beschin et al., 2006 (type)
Pseudohepatiscus†† Blow & Manning, 1996 (0, 0, 2)
P. marinoi Blow & Manning, 1996 (type)
P. silvanoi De Angeli & Beschin, 1999
Incertae sedis
"*Titanocarcinus*" *reisi* Böhm, 1922
Superfamily Leucosioidea Samouelle, 1819
Family **Iphiculidae**† Alcock, 1896
Iphiculus† Adams & White, 1848 (2, 0, 3)
I. granulatus†† Morris & Collins, 1991
I. miriensis†† Morris & Collins, 1991
I. sexspinosus†† Morris & Collins, 1991
Pariphiculus† Alcock, 1896 (3, 0, 6)
P. decemtuberculatus†† Collins et al., 2003
P. gselli†† Beets, 1950
P. inconditus†† Karasawa, 1993
P. multituberculatus†† Collins et al., 2003

P. papillosus†† Morris & Collins, 1991

P. verrucosus†† Morris & Collins, 1991

Leucosiidae† Samouelle, 1819

Subfamily Cryptocneminae Stimpson, 1907

No fossil representatives

Subfamily Ebaliinae Stimpson, 1871a

Arcania† Leach, 1817 [in 1814-1817] (=*Ixoides*† MacGilchrist, 1905) (14, 5, 1)

A. cornuta† (MacGilchrist, 1905)

A. cf. *globata*† Stimpson, 1858b

A. elongata† Yokoya, 1933

A. shuishuni†† (Hu & Tao, 1985)

A. undecimspinosa† De Haan, 1841

A. cf. *undecimspinosa*† De Haan, 1841

Ebalia† Leach, 1817 (64, 7, 12)

E. cranchii† Leach, 1817 [in 1815-1875]

E. cryptocnemoides† Takeda & Miyake, 1972

E. ferganensis†† Ilyin & Alekseev *in* Ilyin, 2005

E. fucinii†† Ristori, 1893

E. hungarica†† Müller, 1974b

E. jacqueshermani†† Van Bakel, Jagt, Fraaije & Wille, 2003

E. kakegawaensis†† Karasawa, 1993

E. lamarmorai†† Lőrenthey, 1909a

E. longimana† Ortmann, 1892b

E. cf. *longimana*† Ortmann, 1892b

E. meulenkampi†† Georgiades & Müller, 1984

E. multiangulata†† Müller, 1993

E. nux† A. Milne-Edwards, 1883

E. oersi†† Müller, 1978

E. tuberosa† (Pennant, 1777) (=*E. pennanti* Leach, 1817 [in 1815-1875])

E. tumefacta† (Montagu, 1808)

E. vahldieki†† Förster & Mundlos, 1992

E. vanstraeleni†† Bachmayer, 1953 [imprint 1951]

E. noetlingi†† Karasawa, 2008 (=*E. tuberculata* Noetling, 1901)

Ilia† Leach, 1817 [in 1814-1817] (2, 1, 2)

  - *I.* cf. *nucleus*† (Linnaeus, 1758)
  - *I. pliocaenica*†† Ristori, 1891
  - *I. valdellae*†† Fabiani, 1908
- *Iliacantha*† Stimpson, 1871a (5, 0, 1)
  - *I. panamanica*†† Collins & Todd *in* Todd & Collins, 2005
- *Heteronucia*† Alcock, 1896 (16, 1, 0)
  - *H. perlata*† (Sakai, 1963)
- *Ixa*† Leach, 1816 [imprint 1815] (8, 1, 0)
  - *I. cylindrus*† (Fabricius, 1777)
- *Leucosilia*† Bell, 1855 (1, 1, 1)
  - *L. bananensis*†† Rathbun, 1918
  - *L. jurinii*† (de Saussure, 1853)
- *Myra*† Leach, 1817 [in 1814-1817] (15, 3, 9)
  - *M.* cf. *affinis*† Bell, 1855
  - *M. amoena*†† Böhm, 1922
  - *M. arachnoides*†† Böhm, 1922
  - *M. brevisulcata*†† Morris & Collins, 1991
  - *M. celeris*† Galil, 2001
  - *M. emarginata*†† Glaessner, 1928
  - *M. ? fijiensis*†† Rathbun, 1934
  - *M. fugax*† (Fabricius, 1798)
  - *M. paenitensis*†† Vía, 1941
  - *M. subcarinata*†† Morris & Collins, 1991
  - *M. trispinosa*†† Morris & Collins, 1991
  - *M. wenjii*†† Hu & Tao, 1985
- *Nucia*† Dana, 1852a (5, 1, 6)
  - *N. calculoides*†† Morris & Collins, 1991
  - *N. coxi*†† Morris & Collins, 1991
  - *N. speciosa*† Dana, 1852a
  - *N.* ? *baripadensis*†† Bachmayer & Mohanti, 1973
  - *N.* ? *borneoensis*†† Morris & Collins, 1991
  - *N.* ? *fennemai*†† Böhm, 1922
  - *N.* ? *platyspinosa*†† Collins et al., 2003
- *Nursia*† Leach, 1817 [in 1814-1817] (19, 1, 1)
  - *N. bilobata*†† Karasawa & Kato, 2008 *in* Karasawa et al., 2008
  - *N.* sp. aff. *N. japonica*† Sakai, 1935
- *Paranursia*† Serène & Soh, 1976 (1, 0, 1)

P. acharyai†† (Bachmayer & Mohanti, 1973)
Persephona† Leach, 1817 [in 1814-1817] (8, 2, 7)
P. bigranulata†† Rathbun, 1926b
P. enigmatica†† Collins & Todd *in* Todd & Collins, 2005
P. invalida†† Rathbun, 1929a
P. manningi†† Collins & Todd *in* Todd & Collins, 2005
P. mediterranea† (Herbst, 1794)
P. niemeyeri†† Blow, 2003
P. prepunctata†† Rathbun, 1919
P. punctata† (Linnaeus, 1758)
P. rodesae†† Blow, 2003
Philyra† Leach, 1817 [in 1814-1817] (39, 6, 12)
P. alveola†† Hu & Tao, 2000
P. cranium†† (Desmarest, 1822)
P. ferrica†† Hu & Tao, 1996
P. granulosa†† Morris & Collins, 1991
P. hayasakai†† Karasawa & Inoue, 1992
P. heterograna† Ortmann, 1892b
P. miyamotoi†† Karasawa & Kishimoto, 1996
P. nishimotoi†† Karasawa, 1989
P. pisum† De Haan, 1841
P. plana†† Karasawa, 1989
P. platycheir† De Haan, 1841
P. scabriuscula† (Fabricius, 1793)
P. shihcheni†† (Hu & Tao, 1979)
P. syndactyla† Ortmann, 1892
P. tanakai†† Karasawa, 1993
P. tridentata†† Karasawa, 1993
P. trusanensis†† Collins et al., 2003
P. sp. cf. *P. acutidens*† Chen, 1987
Pseudophilyra† Miers, 1879 (15, 0, 2)
P. elongatella†† Hu & Tao, 1996
P. granulimarginata†† Karasawa & Kato *in* Karasawa et al., 2008
Randallia† Stimpson, 1857 [1856-1859] [imprint 1859] (8, 1, 5)
R. ornata† (Randall, 1839)
R. pleistocenica†† Rathbun, 1926b

R. strouhali†† Bachmayer, 1953a
R. ? *jingomao*†† Hu & Tao, 2004
R. ? *prolanata*†† Hu & Tao, 1996
R. ? *saitoensis*†† Karasawa, 1993
Speloeophorus† A. Milne-Edwards, 1865b (9, 0, 1)
S. subcircularis†† Collins & Todd *in* Todd & Collins, 2005
Tokoyo† Galil, 2003a (1, 1, 1)
T. eburnea† (Alcock, 1896)
T. trinuclodea†† (Hu & Tao, 1996)
Uhlias† Stimpson, 1871b (1, 1, 0)
U. cf. *limbatus*† Stimpson, 1871b
Subfamily Leucosiinae† Samouelle, 1819
Euclosia† Galil, 2003b (8, 2, 0)
E. obtusifrons† (De Haan, 1841)
E. unidentata† (De Haan, 1841)
Leucosia† Weber, 1795 (23, 2, 11)
L. anatum† (Herbst, 1783)
L. calcarata†† Collins et al., 2003
L. craniolaris† (Linnaeus, 1758)
L. longiangulata†† Morris & Collins, 1991
L. martini†† Karasawa & Kato *in* Karasawa et al., 2008
L. ovalata†† Hu & Tao, 1996
L. serenei†† Morris & Collins, 1991
L. subrhomboidalis†† Desmarest, 1822
L. subrhomboidea†† Hu & Tao, 1996
L. taiwanica†† Hu & Tao, 1996
L. takamii†† Karasawa, 1993
L. tricarinata†† Martin, 1880
L. tutongensis†† Morris & Collins, 1991
Urnalana† Galil, 2005 (18, 1, 0)
U. parhaematostica† Galil, 2005
Fossil **Leucosiidae** genera not assigned to subfamily:
Ampliura†† Morris & Collins, 1991 (0, 0, 1)
A. simplex Morris & Collins, 1991 (type)
Andorina†† Lőrenthey, 1901 (0, 0, 1)
A. elegans Lőrenthey, 1901 (type)
Duncania†† Portell & Collins, 2004 (0, 0, 1)

D. jamaicensis Portell & Collins, 2004 (type)

Gemmacarcinus†† Müller & Collins, 1991 (0, 0, 2)

G. fossatus Müller & Collins, 1991 (type)

G. ? planus Müller, 1993

Hepatinulus†† Ristori, 1886 (0, 0, 2)

H. lovisatoi Lőrenthey, 1909a

H. seguenzai Ristori, 1886 (type)

Nucilobus†† Morris & Collins, 1991 (0, 0, 2)

N. bericus De Angeli & Beschin, 2004

N. symmetricus Morris & Collins, 1991 (type)

Palaeomyra†† A. Milne-Edwards *in* Sismonda, 1861 (0, 0, 2)

P. bispinosa A. Milne-Edwards *in* Sismonda, 1861 (type)

P. globulosa (Müller, 1976b)

Pterocarcinus†† Blow, 2003 (0, 0, 1)

P. baileyi Blow, 2003 (type)

Typilobus†† Stoliczka, 1871 (0, 0, 18)

T. ? belli Quayle & Collins, 1981

T. ? birshteinyi Förster & Mundlos, 1982

T. ? boscoi Vía, 1959

T. ? cf. *corrodatus* (Noetling, 1885a)

T. ? corrodatus (Noetling, 1885a)

T. granulosus Stoliczka, 1871 (type)

T. johnfeni (Hu & Tao, 1996)

T. kishimotoi Karasawa, 1998b

T. marginatus Morris & Collins, 1991

T. ? modregoi Vía, 1959

T. moralejai Müller, 1993

T. ? obscurus Quayle & Collins, 1981

T. ? prevostianus (Desmarest, 1822)

T. sadeki Withers, 1925

T. ? semseyanus Lőrenthey, 1898a

T. ? trispinosus Lőrenthey, 1909b

T. unispinatus Martins-Neto, 2001

Matutidae† De Haan, 1841

Discussion. — We follow Schweitzer & Feldmann (2000b) and Martin & Davis (2001) in placing the Matutidae within the Leucosioidea rather than the Calappoidea as in Ng et al. (2008).

Ashtoret† Galil & Clark, 1994 (8, 0, 1)

Ashtoret sp. (in Karasawa, 2002)

Eomatuta†† De Angeli & Marchiori, 2009 (0, 0, 1)

E. granosa De Angeli & Marchiori, 2009 (type)

Szaboa†† Müller & Galil, 1998 (0, 0, 1)

S. inermis (Brocchi, 1883) (type)

Incertae sedis

"*Necrocarcinus*" *senoniensis* Schlüter *in* Schlüter & Von der Marck, 1868

Superfamily Majoidea Samouelle, 1819

Family **Epialtidae** MacLeay, 1838

Subfamily Epialtinae† MacLeay, 1838

Actinotocarcinus†† Jenkins, 1974 (0, 0, 2)

A. chidgeyorum Jenkins, 1974 (type)

A. maclauchlani Feldmann, 1993

Eoinachoides†† Van Straelen, 1933b (0, 0, 2)

E. latispinosa Carriol, de Muizon & Secretan, 1987

E. senni Van Straelen, 1933b (type)

Epialtus† H. Milne Edwards, 1834 [in 1834-1840] (11, 0, 1)

E. pacificus†† Rathbun, 1945

Nanomaja†† Müller & Collins, 1991 (0, 0, 1)

N. simplex Müller & Collins, 1991 (type)

Panticarcinus†† Collins & Saward, 2006 (0, 0, 1)

P. maylandiensis Collins & Saward, 2006 (type)

Periacanthus†† Bittner, 1875 (0, 0, 4)

P. dallonii Vía, 1959

P. dallagoi Beschin et al., 2005

P. horridus Bittner, 1875 (type)

P. ramosus Artal & Castillo, 2005b

Pugettia† Dana, 1851c (16, 3, 1)

P. incisa† (De Haan, 1837)

P. producta† (Randall, 1839)

P. richii† Dana, 1851c

P. yamafui†† Karasawa & Kitamura, 1993

Subfamily Pisinae† Dana, 1851d

Chorilia† Dana, 1851c (3, 0, 1)

C. yui†† Hu, 1984b

Herbstia† H. Milne Edwards, 1834 [in 1834-1840] (11, 0, 1)
H. exserta†† Collins & Morris, 1976
Hyastenus† White, 1847b (36, 1, 1)
H. corallinus De Angeli & Beschin, 2008
H. diacanthus† (De Haan, 1839)
Libinia† Leach, 1815 [in 1814-1817] (9, 3, 2)
L. dubia† H. Milne Edwards, 1834 [in 1834-1840]
L. emarginata† Leach, 1815 [in 1814-1817]
L. mbalavuensis†† Rathbun, 1945
L. peruviensis†† Carriol, de Muizon & Secretan, 1987
L. setosa† Lockington, 1877c
Loxorhynchus† Stimpson, 1857 [1856-1859] [imprint 1859] (0, 2, 1)
L. crispatus† Stimpson, 1857 [1856-1859] [imprint 1859]
L. grandis† Stimpson, 1857 [1856-1859] [imprint 1859]
L. peruviensis†† Carriol, de Muizon & Secretan, 1987
Pisa† Leach, 1816 [imprint 1815] (8, 1, 2)
P. armata† (Latreille, 1802) (= *P. gibbsi* Leach, 1816 [imprint 1815])
P. oroszyi†† (Bachmayer, 1953 [imprint 1951])
Pisoides† H. Milne Edwards & Lucas, 1843 (2, 2, 0)
P. bidentatus† (A. Milne-Edwards, 1873b)
P. ortmanni† (Balss, 1924)
Pitho† Bell, 1836 (9, 1, 0)
P. anisodon† (Von Martens, 1872)
Rochinia† A. Milne-Edwards, 1875 [1873-1881] (34, 0, 1)
R. boschii†† Casadío et al., 2005
Scyra† Dana, 1851c (2, 1, 0)
S. acutifrons† Dana, 1851c
Tylocarcinus† Miers, 1879 (4, 1, 0)
T. styx† (Herbst, 1803)
Subfamily Pliosomatinae Števčić, 2005
No fossil representatives
Subfamily Tychinae Dana, 1851d
No fossil representatives
Hymenosomatidae MacLeay, 1838

No fossil representatives

Inachidae† MacLeay, 1838

Achaeus† Leach, 1817 [in 1815-1875] (36, 0, 2)

A. magnus†† Müller, 1978

A. nomurai†† Takeda & Fujiyama, 1984

Inachus† Weber, 1795 (12, 1, 1)

I. lethaeus†† Smirnov, 1929

I. phalangium† (Fabricius, 1775) [=*I. dorhynchus* Leach, 1814]

Macrocheira† De Haan, 1839 (1, 0, 4)

M. ginzanensis†† Imaizumi, 1965

M. longirostra†† Schweitzer & Feldmann, 1999

M. teglandi†† Rathbun, 1926b

M. yabei†† (Imaizumi, 1957b)

Stenorhynchus† Lamarck, 1818 (4, 0, 1)

S. lethaeus†† Smirnov, 1929

Inachoididae† Dana, 1851d

Euprognatha† Stimpson, 1871a (6, 0, 1)

E. ricei†† Blow, 2003

Leurocyclus† Rathbun, 1897a (2, 0, 1)

L. primigenius†† Aguirre-Urreta, 1990

Pyromaia† Stimpson, 1871a (5, 1, 1)

P. inflata†† Collins & Morris, 1978

P. tuberculata† Lockington, 1877a

Majidae† Samouelle, 1819

Subfamily Eurynolambrinae Števčić, 2005

No fossil representatives

Subfamily Majinae† Samouelle, 1819

Jacquinotia† Rathbun, 1915 (0, 1, 0)

J. edwardsi† (Jacquinot & Lucas, 1853)

Leptomithrax† Miers, 1876 (13, 1, 8)

L. atavus†† Glaessner, 1960

L. edwardsii† (De Haan, 1837)

L. elegans†† Jenkins, 1985

L. elongatus†† McLay et al., 1995

L. garthi†† McLay et al., 1995

L. griffini†† Feldmann & Maxwell, 1990

L. irirangi†† Glaessner, 1960

L. martensis†† Jenkins, 1985
L. uruti†† Glaessner, 1960
Maja† Lamarck, 1801 (15, 2, 12)
M. arambourgi†† Van Straelen, 1936d
M. austriaca†† Bachmayer, 1953 [imprint 1951]
M. biaensis†† Lőrenthey *in* Lőrenthey & Beurlen, 1929
M. crispata† Risso, 1826 (=*M. verrucosa* H. Milne Edwards, 1834 [in 1834-1840])
M. dominoleuae†† Hu & Tao, 1985
M. hungarica†† Lőrenthey *in* Lőrenthey & Beurlen, 1929
M. laddi†† Rathbun, 1934
M. miocaenica†† Lőrenthey, 1907a
M. morii†† Kato, 1996
M. orbigniana†† Millet, 1854
M. oroszyi†† Bachmayer, 1953 [imprint 1951]
M. robinsoni†† Jenkins, 1985
M. squinado† (Herbst, 1788)
M. tomidai†† Kato, 2002
Notomithrax† Griffin, 1963 (4, 0, 2)
N. allani†† Feldmann & Maxwell, 1990
N. angustifrons†† Jenkins, 1985
Schizophroida† Sakai, 1933 (2, 0, 1)
S. tertiaria†† Jenkins, 1985
Schizophrys† White, 1847a (6, 0, 1)
S. visegradensis†† Müller, 1984a
Wilsonimaia†† Blow & Manning, 1996 (0, 0, 2)
W. ethelae Blow & Manning, 1996 (type)
W. schneiderorum Blow & Manning, 1996
Subfamily Micromaiinae†† Beurlen, 1930a
Micromaia†† Bittner, 1875 (0, 0, 11)
M. batalleri Vía, 1959
M. elegans Beschin et al., 1985
M. laevis Lőrenthey, 1909b
M. gulderritteri Bachmayer, 1950
M. mainensis Beschin et al., 1985
M. margaritata Fabiani, 1910
M. meneguzzoi Beschin et al., 1985

M. priabonensis Oppenheim, 1901

M. punctulata Lőrenthey *in* Lőrenthey & Beurlen, 1929

M. spinosa Noetling, 1885a

M. simplex Remy *in* Gorodiski & Remy, 1959

M. tuberculata Bittner, 1875 (type)

Mithracia†† Bell, 1858 (0, 0, 4)

M. hollanda Förster & Mundlos, 1982

M. libinioides Bell, 1858 (type)

M. margaritifera Beschin et al., 1994

M. oppioni Larghi, 2002

Pisomaja†† Lőrenthey *in* Lőrenthey & Beurlen, 1929 (0, 0, 1)

P. tuberculata Lőrenthey *in* Lőrenthey & Beurlen, 1929 (type)

Subfamily Mithracinae† MacLeay, 1838

Antarctomithrax†† Feldmann, 1994 (0, 0, 1)

A. thomsoni Feldmann, 1994 (type)

Micippa† Leach, 1817 [in 1814-1817] (9, 1, 2)

M. antiqua†† Beschin, De Angeli & Checchi, 2001

M. hungarica†† (Lőrenthey *in* Lőrenthey & Beurlen, 1929)

M. thalia† (Herbst, 1803)

Mithraculus† White, 1847a (7, 2, 0)

M. cf. *M. forceps*† A. Milne-Edwards, 1875 [in 1873-1881]

M. ruber† Stimpson, 1871a

Mithrax† Latreille, 1817a (12, 4, 2)

M. donovani†† Portell & Collins, 2004

M. hemphilli† Rathbun, 1892

M. hispidus† (Herbst, 1790) [=*M. caribbaeus*† Rathbun, 1920b]

M. spinosissimus† (Lamarck, 1818)

M. unguis†† Portell & Collins, 2004

M. verrucosus† H. Milne Edwards, 1832

Nemausa† A. Milne-Edwards, 1875 [in 1873-1881] (2, 1, 0)

N. acuticornis† (Stimpson, 1870 [in 1870-1881])

Stenocionops† Desmarest, 1823 (8, 0, 3)

S. dyeri†† Blow, 2003
S. primus†† Rathbun, 1935a
S. suwanneeana†† Rathbun, 1935a
Teleophrys† Stimpson, 1860 (4, 0, 1)
T. acornis†† Portell & Collins, 2004
Thoe† Bell, 1836 (4, 0, 2)
T. asperoides†† Collins & Todd *in* Todd & Collins, 2005
T. vanuaensis†† Rathbun, 1945
Subfamily Planoterginae Števčić, 2005
No fossil representatives
Oregoniidae† Garth, 1958
Hyas† Leach, 1814 (4, 1, 4)
H. aranaeus† (Linnaeus, 1758)
H. meridionalis†† Glaessner, 1928
H. oranensis†† Van Straelen, 1936d
H. tani†† Hu, 1984b
H. tsuchidai†† Imaizumi, 1952d
Oregonia† Dana, 1851c (2, 0, 1)
O. spinifera†† Schweitzer, Feldmann, Gonzáles-Barba & Vega, 2002
Superfamily Parthenopoidea MacLeay, 1838
Family **Parthenopidae**† MacLeay, 1838
Subfamily Daldorfiinae† Ng & Rodriguez, 1986
Daldorfia† Rathbun, 1904a (12, 0, 5)
D. eocenica†† Beschin, Busulini, De Angeli & Tessier, 2007
D. fabianii†† Beschin, De Angeli & Checchi, 2001
D. himaleoraphis†† Schweitzer, 2001
D. nagashimai†† Karasawa & Kato, 1996
D. salina†† Schweitzer, Feldmann, Gonzáles-Barba & Ćosović, 2006
Subfamily Parthenopinae† MacLeay, 1838
Acantholambrus†† Blow & Manning, 1996 (0, 0, 1)
A. baumi Blow & Manning, 1996 (type)
Bittnerilia†† De Angeli & Garassino, 2003 (0, 0, 1)
B. eocaena†† (Bittner, 1883) (type)
Derilambrus† Tan & Ng, 2007 (0, 1, 0)
D. angulifrons† (Latreille, 1825)

Enoplolambrus† A. Milne-Edwards, 1878 [in 1873-1881] (6, 1, 0)

- *E. laciniata*† (De Haan, 1839)

Latolambrus† Tan & Ng, 2007 (0, 1, 0)

- *L. occidentalis*† (Dana, 1854)

Leiolambrus† A. Milne-Edwards, 1879 [in 1873-1881] (2, 0, 1)

- *L. longifrons*†† Quayle & Collins, 1981

Mesorhoea† Stimpson, 1871a (1, 1, 2)

- *M. idae*†† Rathbun, 1926b
- *M. mauryae*†† Rathbun, 1919
- *M. sexspinosa*† Stimpson, 1871a

Parthenope† Weber, 1795 (7, 0, 13)

- *P. bachmayeri*†† Förster & Mundlos, 1982
- *P. bittneri*†† (Noetling, 1885a)
- *P. clypeformis*†† Collins et al., 2003
- *P. jugular*†† Collins et al., 2003
- *P. neocomiensis*†† Robineau-Desvoidy, 1849
- *P. nummulitica*†† (Bittner, 1875)
- *P. panamensis*†† Rathbun, 1918
- *P. pleistocenica*†† Rathbun, 1918
- *P. prisca*†† (A. Milne-Edwards, 1873a)
- *P. szaboi*†† Müller, 1974a
- *P. tetenyensis*†† Müller, 1984a
- *P. trituberculata*†† Beurlen, 1958a
- *P. venezuelensis*†† Van Straelen, 1933b

Platylambrus† Stimpson, 1871a (2, 0, 4)

- *P. charlottensis*†† Rathbun, 1935a
- *P. fijiensis*†† Rathbun, 1934
- *P. obscura*†† Rathbun, 1919
- *P. spinulatus*†† Collins & Todd *in* Todd & Collins, 2005

Phrynolambrus†† Bittner, 1893 (0, 0, 4)

- *P. corallinus* (Bittner, 1893) (type)
- *P. italicus* De Angeli & Beschin, 2008
- *P. pentagonus* Quayle & Collins, 1981
- *P. weinfurteri* Bachmayer, 1953 [imprint 1951]

Rhinolambrus† A. Milne-Edwards, 1879 [in 1873-1881] (12, 1, 1)

R. pelagicus† (Rüppell, 1830)
R. sublitoralis†† Morris & Collins, 1991
Tutankhamen† Rathbun, 1925 (1, 0, 1)
T. hieracodes†† Jenkins, 1985
Fossil **Parthenopidae** not assigned to subfamiily
Branchiolambrus†† Rathbun, 1908 (0, 0, 1)
B. altus Rathbun, 1908 (type)
Mesolambrus†† Müller & Collins, 1991 (0, 0, 1)
M. declinatus Müller & Collins, 1991 (type)
Superfamily Retroplumoidea Gill, 1894
Family **Retroplumidae**† Gill, 1894
Archaeopus†† Rathbun, 1908 (0, 0, 8)
A. antennatus Rathbun, 1908 (type)
A. bicornutus Schweitzer, Feldmann et al., 2003
A. ezoensis (Nagao, 1941)
A. lunicarinus Schweitzer & Feldmann, 2001a
A. mexicanus Schweitzer, Feldmann et al., 2002
A. rostratus Schweitzer, Feldmann et al., 2003
A. schenki (Van Straelen, 1939a)
A. vancouverensis (Woodward, 1896)
Costacopluma†† Collins & Morris, 1975 (0, 0, 10)
C. australis Feldmann, Casadío et al., 1995
C. bifida Collins et al., 1994
C. binodosa Collins & Rasmussen, 1992
C. bishopi Vega & Feldmann, 1992
C. concava Collins & Morris, 1975 (type)
C. grayi Feldmann & Portell, 2007
C. mexicana Vega & Perrilliat, 1989
C. nordestina Feldmann & Martins-Neto, 1995
C. salamanca Feldmann et al., 1997
C. senegalensis (Rémy *in* Gorodiski & Rémy, 1959)
Cristipluma†† Bishop, 1983c (0, 0, 1)
C. mississippiensis Bishop, 1983c (type)
Loerentheya†† Beurlen *in* Lőrenthey & Beurlen, 1929 (0, 0, 1)
L. carinata Beurlen *in* Lőrenthey & Beurlen, 1929 (type)
Loerenthopluma†† Beschin, Busulini, De Angeli & Tessier, 1996a (0, 0, 1)

L. lata Beschin et al., 1996a (type)
Retrocypoda†† Vía, 1959 (0, 0, 1)
R. almelai Vía, 1959 (type)
Retropluma† Gill, 1894 (7, 0, 5)
R. borealis†† Fraaije, Hansen & Hansen, 2005
R. craverii†† (Crema, 1895)
R. eocenica†† Vía, 1959
R. gallica†† Artal, Van Bakel & Castillo, 2006
R. laurentae†† Collins et al., 2003
Superfamily Cancroidea Latreille, 1802
Family **Atelecyclidae**† Ortmann, 1893
Atelecyclus† Leach, 1814 (2, 0, 3)
A. elegans†† Ristori, 1896
A. rotundatus†† (Olivi, 1792)
A. rugosus†† Desmarest, 1822
Levicyclus†† Schweitzer, Feldmann, González-Barba & Vega, 2002 (0, 0, 1)
L. tepetate Schweitzer et al., 2002 (type)
Palaeotrichia†† Guinot, 1976 (0, 0, 2)
P. laevis (Noetling, 1885a) (type)
P. multispinata (Noetling, 1885a)
Podocatactes† Ortmann, 1893 (0, 1, 1)
P. hamifer† Ortmann, 1893 (type)
P. ujiharai†† Karasawa, 1993
Trichopeltarion† A. Milne-Edwards, 1880a (=*Trachycarcinus* Faxon, 1893) (18, 0, 8)
T. berglundorum†† Schweitzer & Feldmann, 1999
T. decorus†† (Rathbun, 1945)
T. granulosa†† Schweitzer & Salva, 2000
T. greggi†† Dell, 1969
T. huziokai†† (Imaizumi, 1951)
T. inflatus†† (Kato, 1996)
T. levis†† Casadío et al., 2004
T. merrinae†† Schweitzer & Salva, 2000
Cancridae† Latreille, 1802
Subfamily Cancrinae Latreille, 1802
Anatolikos† Schweitzer & Feldmann, 2000c (1, 1, 2)
A. japonicus† (Ortmann, 1893) (type)

A. itoigawai†† (Karasawa, 1990)
A. undecimspinosus†† Schweitzer, Feldmann, González-Barba & V. Ćosović, 2006
Anisospinos†† Schweitzer & Feldmann, 2000c (0, 0, 3)
A. berglundi Schweitzer & Feldmann, 2000c
A. odosensis (Imaizumi, 1962)
A. wahkiakumensis (Berglund & Goedert, 1992) (type)
Cancer† Linnaeus, 1758 (1, 7, 26)
C. allisoni†† Nations, 1975
C. bainbridgensis†† Rathbun, 1926b
C. bellianus† Johnson, 1861
C. borealis† Stimpson, 1862a
C. burtini†† Galeotti, 1837
C. chaneyi†† Nations, 1975
C. coosensis†† Nations, 1975
C. durhami†† Nations, 1975
C. fissus†† Rathbun, 1908
C. fujinaensis†† Sakumoto, Karasawa & Takayasu, 1992
C. gabbi†† Rathbun, 1926b
C. garthi†† Nations, 1975
C. illyricus†† Bittner, 1883
C. irroratus† Say, 1817
C. johngarthi† Carvacho, 1989
C. marri†† Nations, 1975
C. meticurensis†† Thurmann, 1853
C. modestus†† Fritsch & Kafka, 1887
C. pagurus† Linnaeus, 1758
C. parvidens†† Collins & Fraaye, 1991
C. porteri† Rathbun, 1930c
C. proavitus†† Packard, 1900
C. productus† Randall, 1839
C. reversus†† Fritsch & Kafka, 1887
C. santosi†† (Rathbun, 1937)
C. solitarius†† Fritsch & Kafka, 1887
C. spinosus†† (Ristori, 1886)
C. styriacus†† Bittner, 1884
C. tomowoi†† Karasawa, 1990
C. vancalsteri†† Van Bakel, Jagt & Fraaije, 2003

C. villabersiani†† Gary, 1907
C. whitfieldi†† Pilsbry, 1901
C. yanceyi†† Nations, 1975
Ceronnectes†† De Angeli & Beschin, 1998 (0, 0, 3)
C. boeckhi (Lőrenthey, 1898a) (type)
C. granulosa (Feldmann, Bice et al., 1998)
? *C. pusillinus* (Secretan *in* Plaziat & Secretan, 1971)
Cyclocancer†† Beurlen, 1958a (0, 0, 1)
C. tuberculatus Beurlen, 1958a (type)
Glebocarcinus† Nations, 1975 (1, 1, 0)
G. amphioetus† (Rathbun, 1898a)
Metacarcinus† A. Milne-Edwards, 1862c (1, 4, 9)
M. anthonyi† (Rathbun, 1897b)
M. danai†† Nations, 1975
M. davidi†† (Nations, 1968)
M. goederti†† Schweitzer & Feldmann, 2000c
M. gracilis† (Dana, 1852a)
M. granti†† (Rathbun, 1932b)
M. izumoensis†† Sakumoto, Karasawa & Takayasu, 1992
M. jenniferae†† Nations, 1975
M. magister† (Dana, 1852a)
M. minutoserratus†† (Nagao, 1940)
M. novaezelandiae† (Jacquinot *in* Jacquinot & Lucas, 1853)
M. starri†† Berglund & Goedert, 1996
M. tenax†† Van Bakel, Jagt, Fraaije & Wille, 2003
Microdium†† Reuss, 1867 (0, 0, 1)
M. nodulosum Reuss, 1867 (type)
Notocarcinus†† Schweitzer & Feldmann, 2000c (0, 0, 1)
N. sulcatus Schweitzer & Feldmann, 2000c (type)
Platepistoma† Rathbun, 1906 (7, 0, 2)
P. imamurae†† (Imaizumi, 1962)
P. kaedei†† (Karasawa, 1992a)
Romaleon† Gistl, 1848 (3, 4, 7)
R. antennarium† (Stimpson, 1873)
R. branneri†† (Rathbun, 1926b)
R. dereki†† Nations, 1975

R. gibbosulum† (De Haan, 1833) (type)
R. jordani† (Rathbun, 1900)
R. nakagawaensis†† Kato & Hikida, 2002
R. parspinosus†† Casadío et al., 2005
R. polyodon† (Poeppig, 1836)
R. sakamotoi†† (Kato, 1996)
R. sanbonsugii†† (Imaizumi, 1962)
R. urbanus†† (Rathbun, 1917b)
Santeecarcinus†† Blow & Manning, 1996 (0, 0, 1)
S. harmatuki Blow & Manning, 1996 (type)
Sarahcarcinus†† Blow & Manning, 1996 (0, 0, 1)
S. campbellorum Blow & Manning, 1996 (type)
Subfamily Lobocarcininae†† Beurlen, 1930a
Lobocarcinus†† Reuss, 1857 (0, 0, 8)
L. aegypticus Lőrenthey, 1909b
L. catalaunicus Van Straelen, 1927
L. indicus Glaessner, 1933b
L. lumacopius Anderson & Feldmann, 1995
L. paulinowuerttemburgensis Von Meyer, 1847c (type)
L. pentanodosus Collins & Barber, 1998
L. pustulosus Feldmann & Fordyce, 1996
L. sismondai Von Meyer, 1843
Miocyclus†† Müller, 1979 (0, 0, 1)
M. bulgaricus Müller, 1979 (type)
Tasadia†† Müller *in* Janssen & Müller, 1984 (0, 0, 1)
T. carniolica (Bittner, 1884) (type) (= *Cancer szontaghii* Lőrenthey, 1898a; *C. bittneri*†† Toula, 1904)
Cheiragonidae† Ortmann, 1893
Karasawaia†† Vega et al., 2008 (0, 0, 1)
K. markgrafi (Lőrenthey, 1907b [German version 1909b]) (type)
Montezumella†† Rathbun, 1930b (0, 0, 11)
M. amenosi Vía, 1959
M. casayetensis Rathbun, 1937
M. eichhorni Schweitzer & Salva, 2000
M. elegans (Lőrenthey *in* Lőrenthey & Beurlen, 1929)
M. fraasi (Lőrenthey, 1909b)
M. lamiensis Rathbun, 1934

M. microporosa Portell & Collins, 2002
M. pumicosa Beschin et al., 2002
M. rutteni Van Straelen, 1933c
M. scabra Quayle & Collins, 1981
M. tubulata Rathbun, 1930b (type)
Stintonius†† Collins, 2002 (0, 0, 1)
S. subovata (Quayle & Collins, 1981) (type)
Incertae sedis
"*Xanthilites*" *rathbunae* Withers, 1924a
Corystidae† Samouelle, 1819
Corystes† Latreille, 1802 (0, 1, 1)
C. cassivelaunus† (Pennant, 1777) (type)
C. minisculus†† (Feldmann & Wilson, 1988)
Corystites†† Lőrenthey *in* Lőrenthey & Beurlen, 1929 (type is lost) (0, 0, 1)
C. latifrons Lőrenthey *in* Lőrenthey & Beurlen, 1929 (type)
Gomezinus†† Collins, Lee & Noad, 2003 (0, 0, 1)
G. tuberculatus Collins et al., 2003 (type)
Harenacorystes†† Van Bakel et al., 2009 (0, 0, 1)
H. johnjansseni†† Van Bakel et al., 2009 (type)
Hebertides†† Guinot, De Angeli & Garassino, 2007 (0, 0, 1)
H. jurassica Guinot, De Angeli & Garassino, 2007 (type)
Micromithrax†† Noetling, 1881 (0, 0, 1)
M. holsatica Noetling, 1881 (type)
Superfamily Dairoidea Serène, 1965
Family **Dacryopilumnidae** Serène, 1984
No fossil representatives
Dairidae† Ng & Rodriguez, 1986
Daira† De Haan, 1833 (1, 1, 8)
D. coronata†† Beschin et al., 2005
D. depressa†† (A. Milne-Edwards, 1865c)
D. eocaenica†† (Lőrenthey, 1898a)
D. eocenica†† (Di Salvo, 1933)
D. perlata† (Herbst, 1790)
D. salebrosa†† Beschin et al., 2002

D. sicula†† (Di Salvo, 1933)
D. speciosa†† Reuss, 1871
D. vulgaris†† Portell & Collins, 2004
Superfamily Portunoidea Rafinesque, 1815
Family **Carcineretidae**†† Beurlen, 1930a
Branchiocarcinus†† Vega, Feldmann & Sour-Tovar, 1995 (0, 0, 1)
B. cornatus Vega et al., 1995 (type)
Cancrixantho†† Van Straelen, 1934 (0, 0, 1)
C. pyrenaicus Van Straelen, 1934 (type)
Carcineretes†† Withers, 1922 (0, 0, 2)
C. woolacotti Withers, 1922 (type)
C. planetarius Vega et al., 1997
Mascaranada†† Vega & Feldmann, 1991 (0, 0, 1)
M. difuntaensis Vega & Feldmann, 1991 (type)
Carcinidae† MacLeay, 1838
Subfamily Carcininae† MacLeay, 1838
Carcinus† Leach, 1814 (1, 1, 1)
C. maenus† (Linnaeus, 1767)
C. minor†† Rathbun, 1926b
Cicarnus†† Karasawa & Fudouji, 2000 (0, 0, 1)
C. fumiae Karasawa & Fudouji, 2000 (type)
Miopipus†† Müller, 1984a (0, 0, 1)
M. pygmeus (Brocchi, 1883) (type)
Portumnus† Leach, 1814 (3, 0, 1)
? *P. tricarinatus*†† Lőrenthey *in* Lőrenthey & Beurlen, 1929
Xaiva† MacLeay, 1838 (3, 0, 1)
X. bachmayeri†† Müller, 1984a
Subfamily Polybiinae† Paul'son, 1875
Liocarcinus† Stimpson, 1871a (7, 5, 7)
L. atropatanus†† (Aslanova & Dschafarova, 1975)
L. corrugatus† (Pennant, 1777)
L. depurator† (Linnaeus, 1758)
L. holsatus† (Fabricius, 1798)
L. kuehni†† (Bachmayer, 1953 [imprint 1951])
L. lancetidactylus†† (Smirnov, 1929)
L. marmoreus† (Leach, 1816 [imprint 1815])

L. oligocaenicus†† (Paucă, 1929) (= *Portunus musceli* Paucă, 1929)
L. oroszyi†† (Bachmayer, 1953 [imprint 1851])
L. praearcuatus†† Müller, 1996
L. pusillus† (Leach, 1816 [imprint 1815])
L. rakosensis†† (Lőrenthey *in* Lőrenthey & Beurlen, 1929)
Ovalipes† Rathbun, 1898a (10, 1, 1)
O. formosanus†† Hu & Tao, 1996
O. punctatus† (De Haan, 1833)
Catoptridae† Borradaile, 1902
Libystes† A. Milne-Edwards, 1867 (= *Carcinoplacoides* Kesling, 1958) (5, 1, 0)
L. nitidus† A. Milne-Edwards, 1867
Geryonidae† Colosi, 1923
Archaeogeryon†† Colosi, 1923 (0, 0, 1)
A. fuegianus Colosi, 1923 (type)
Archaeoplax†† Stimpson, 1863 (0, 0, 1)
A. signifera Stimpson, 1863 (type)
Chaceon† Manning & Holthuis, 1989 (27, 0, 6)
C. heimertingensis†† (Bachmayer & Wagner, 1957)
C. helmstedtense†† (Bachmayer & Mundlos, 1968)
C. matsushitai†† Kato & Koizumi, 2001
C. miocenicus†† Fraaije, Hansen & Hansen, 2005
C. ottnangensis†† (Bachmayer, 1953a)
C. peruvianus†† (d'Orbigny, 1842)
Lithophylacidae Van Straelen, 1936a
Lithophylax†† A. Milne-Edwards & Brocchi, 1879 (0, 0, 1)
L. trigeri A. Milne-Edwards & Brocchi, 1879 (type)
Longusorbiidae†† Karasawa, Schweitzer & Feldmann, 2008
Longusorbis†† Richards, 1975 (0, 0, 3)
L. cuniculosus Richards, 1975 (type)
L. quadratus Fraaije, Vega, Van Bakel & Garibay-Romero, 2006
L. eutychius Schweitzer, Feldmann & Karasawa, 2007
Macropipidae† Stephenson & Campbell, 1960
Boschettia†† Busulini et al., 2003 (0, 0, 1)

B. giampietroi Busulini et al., 2003 (type)
Falsiportunites†† Collins & Jakobsen, 2003 (0, 0, 1)
F. longispinosus Collins & Jakobsen, 2003 (type)
Gecchelicarcinus†† Beschin, Busulini, De Angeli & Tessier, 2007 (0, 0, 1)
G. lorigae Beschin et al., 2007 (type)
Macropipus† Prestandrea, 1833 (3, 1, 2)
M. gruneri†† Bachmayer & Rutsch, 1962
? *M. ovalipes*†† Secretan, 1975
M. tuberculatus† (Roux, 1828)
Maeandricampus†† Schweitzer & Feldmann, 2002 (0, 0, 2)
M. triangulum (Rathbun, 1926b) (type)
M. granuliferum (Glaessner, 1960)
Megokkos†† Schweitzer & Feldmann, 2000d (0, 0, 4)
M. alaskensis (Rathbun, 1926b) (type)
M. feldmanni (Nyborg et al., 2003)
M. hexagonalis (Nagao, 1932)
M. macrospinus (Schweitzer et al., 2000)
Minohellenus†† Karasawa, 1990 (= *Imaizumila* Karasawa, 1993; *Itoigawaia* Karasawa et al., 1992) (0, 0, 8)
M. araucanus (Philippi, 1887)
M. chichibuensis Kato, 1996
M. inexpressus Schweitzer & Feldmann, 2002
M. macrocheilus Kato & Karasawa, 1994
M. minoensis (Karasawa, 1990)
M. quinquedentatus Karasawa, 1990 (type)
M. sexdentatus (Karasawa, 1993)
M. umemotoi (Karasawa, 1993)
Necora† Holthuis, 1987 (0, 1, 0)
N. puber† (Linnaeus, 1767)
Ophthalmoplax†† Rathbun, 1935a (0, 0, 5)
O. brasiliana (Maury, 1930)
O. comancheensis Rathbun, 1935a
O. stephensoni Rathbun, 1935a (type)
O. ? *spinosus* Feldmann et al., 1999
O. triambonatus Feldmann & Villamil, 2002
Parathranites† Miers, 1886 (8, 0, 1)

P. shibatai†† Karasawa, 1990
Pleolobites†† Remy, 1960 (0, 0, 1)
P. erinaceus Remy, 1960 (type)
Pororaria†† Glaessner, 1980 (0, 0, 1)
P. eocenica Glaessner, 1980 (type)
Portufuria†† Collins, Schulz & Jakobsen, 2005 (0, 0, 1)
P. enigmatica Collins et al., 2005 (type)
Portunites†† Bell, 1858 (0, 0, 9)
P. angustata Collins, Moody & Sandman, 1999
P. eocenica Lőrenthey *in* Lőrenthey & Beurlen, 1929
P. incerta Bell, 1858 (type)
P. insculpta Rathbun, 1926b
P. kattachiensis Karasawa, 1992a
P. nodosa Schweitzer & Feldmann, 2000d
P. rosenfeldi De Angeli & Garassino, 2006a
P. stintoni Quayle, 1984
P. sylviae Quayle & Collins, 1981
Proterocarcinus†† Feldmann et al., 1995 (0, 0, 4)
P. latus (Glaessner, 1933b)
P. lophos Feldmann et al., 1995 (type)
P. corsolini Casadío et al., 2004
P. navidad Feldmann et al., 2005
Rhachiosoma†† Woodward, 1871 (0, 0, 2)
R. bispinosa Woodward, 1871 (type)
R. echinata Woodward, 1871
Pirimelidae† Alcock, 1899
Parapirimela†† Van Straelen, 1937 (0, 0, 1)
P. angolensis Van Straelen, 1937 (type)
Pirimela† Leach, 1816 [in 1815-1875] (1, 0, 1)
P. lorentheyi†† (Müller, 1974b)
Pliopirimela†† Van Bakel, Jagt, Fraaije & Wille, 2003 (0, 0, 1)
P. deconincki Van Bakel et al., 2003 (type)
Trachypirimela†† Müller, 1974a (0, 0, 1)
T. grippi (Müller, 1974a) (type)
Portunidae† Rafinesque, 1815
Enoplonotus†† A. Milne-Edwards, 1860 (0, 0, 1)
E. armatus A. Milne-Edwards, 1860 (type)

Subfamily Atoportuninae† Števčić, 2005

Euronectes†† Karasawa, Schweitzer & Feldmann, 2008 (0, 0, 2)

E. grumiensis (Beschin, De Angeli & Checchi, 2001) (type)

E. vocans (Müller, 1993)

Laleonectes† Manning & Chace, 1990 (2, 1, 0)

L. vocans† (A. Milne-Edwards, 1878b)

Subfamily Caphyrinae† Paul'son, 1875

Lissocarcinus† Adams & White, 1848 (9, 0, 1)

L. szoeraenyiae†† (Müller, 1974a)

Mioxaiva†† Müller, 1978 (0, 0, 1)

M. psammophila Müller, 1978 (type)

Subfamily Carupinae† Paul'son, 1875

Carupa† Dana, 1851b (1, 1, 0)

C. tenuipes† Dana, 1852c = *C. laeviuscula* Heller, 1861 *in* Hu & Tao, 1996

Neptocarcinus†† Lőrenthey, 1898a (0, 0, 1)

N. millenarus Lőrenthey, 1898a (type)

Rakosia†† Müller, 1984a (0, 0, 2)

R. carupoides Müller, 1984a (type)

R. rectifrons Müller, 1996

Subfamily Necronectinae† Glaessner, 1928

Necronectes†† A. Milne-Edwards, 1881b (0, 0, 13)

N. batalleri (Vía, 1932)

N. beaumonti (A. Milne-Edwards, 1864)

N. catalaunicus (Vía, 1941)

N. collinsi Schweitzer, M. Iturralde-Vinent, Hetler & Velez-Juarbe, 2006

N. drydeni Rathbun, 1935a

N. iranensis Toraby & Yazdi, 2002 (*nomen nudum*, published in an abstract with no description or illustrations)

N. nodosus Schweitzer, Feldmann et al., 2002

N. proavitus (Rathbun, 1918)

N. schafferi Glaessner, 1928

N. summus Collins & Donovan, 1995

N. tajinensis Vega et al., 1999

N. vicksburgensis (Stenzel, 1935) (= *N. vaughni* Rathbun, 1935a)
N. vidalianus A. Milne-Edwards, 1881b (type)
Scylla† De Haan, 1833 (3, 1, 9)
S. costata†† Rathbun, 1919
S. floridana†† Rathbun, 1935a
S. hassiaca†† Ebert, 1887
S. laevis†† Böhm, 1922
S. marianae†† Hu & Tao, 1996
S. michelini†† A. Milne-Edwards, 1860
S. molassica†† Studer, 1898
S. ozawai†† Glaessner, 1933b
S. serrata† (Forskål, 1775)
S. sindensis†† (Stoliczka, 1871)
Subfamily Lupocyclinae† Paul'son, 1875
Lupocyclus† Adams & White, 1848 (6, 0, 1)
L. tuberculosus†† Karasawa, 1993
Subfamily Podophthalminae† Dana, 1851b
Euphylax† Stimpson, 1862b (2, 0, 6)
E. callinectias†† Rathbun, 1918
E. domingensis†† (Rathbun, 1919)
E. fortis†† Rathbun, 1918
E. fortispinosus†† Collins et al., 2001
E. maculatus†† Collins & Todd *in* Todd & Collins, 2005
E. septendentatus†† Beurlen, 1958a
Podophthalmus† Lamarck, 1801 (2, 1, 3)
P. defrancei Desmarest, 1822
P. fusiformis†† Morris & Collins, 1991
P. taiwanicus†† Hu & Tao, 1985
P. vigil† (Fabricius, 1798) (type)
Psygmophthalmus†† Schweitzer, M. Iturralde-Vinent, Hetler & Velez-Juarbe et al., 2006 (0, 0, 2)
P. bifurcatus Collins *in* Collins et al., 2009
P. lares Schweitzer et al., 2006 (type)
Sandomingia†† Rathbun, 1919 (0, 0, 1)
S. yaquiensis Rathbun, 1919 (type)
Saratunus†† Collins et al., 2003 (0, 0, 2)
S. longiorbis Collins et al., 2003 (type)

S. karasawai (Collins et al., 2003)
Viaophthalmus†† Karasawa, Schweitzer & Feldmann, 2008 (0, 0, 1)
V. zariqueyi (Vía, 1959) (type)
Subfamily Portuninae† Rafinesque, 1815
Acanthoportunus†† Schweitzer & Feldmann, 2002 (0, 0, 1)
A. buchanani Schweitzer & Feldmann, 2002 (type)
Arenaeus† Dana, 1851b (1, 1, 0)
A. cribrarius† (Lamarck, 1818)
Callinectes† Stimpson, 1862b (13, 3, 6)
C. alabamensis†† Rathbun, 1935a (claw only)
C. bellicosus† (Stimpson, 1862a)
C. declivis†† Rathbun, 1918 (fingers only)
C. jamaicensis†† Withers, 1924a (claw fragment only)
C. paraensis†† Beurlen, 1958a
C. quilonensis†† Verma, 1977
C. reticulatus†† Rathbun, 1918 (claws only)
C. sapidus† Rathbun, 1896 (type)
C. toxodes† Ordway, 1863
Colneptunus†† Lőrenthey *in* Lőrenthey & Beurlen, 1929 (=*Allogoneplax* Van Straelen *in* Dalloni, 1930) (0, 0, 3)
C. hungaricus Lőrenthey *in* Lőrenthey & Beurlen, 1929 (type)
C. hungaricus lutetianus Remy *in* Remy & Tessier, 1954
C. dalloni (Van Straelen *in* Dalloni, 1930)
Cronius† Stimpson, 1862b (2, 0, 1)
C. obscurus†† Rathbun, 1945 (claw only)
Portunus† Weber, 1795 (92, 6, 42) (subgenera not recognized for fossil species)
P. alphonsei†† Karasawa, Schweitzer & Feldmann, 2008 (=*P. granulatus* A. Milne-Edwards, 1860)
P. arcuatus†† (A. Milne-Edwards, 1860)
P. atecuicitlis†† Vega et al., 1999
P. brouweri†† Van Straelen, 1924
P. catalanicus†† (Vía, 1941)

P. delgadoi†† Fontannes, 1884
P. edwardsi†† Sismonda, 1861
P. gabbi†† Rathbun, 1919
P. gallicus†† (A. Milne-Edwards *in* Bouillé, 1873)
P. gibbesii† (Stimpson, 1862a)
P. haitensis†† Rathbun, 1923b
P. hastatus† (Linnaeus, 1767)
P. incertus†† (A. Milne-Edwards, 1860)
P. kisslingi†† Studer, 1892
P. kochi†† (Bittner, 1893)
P. krambergeri†† Bittner, 1893
P. larteti†† (A. Milne-Edwards, 1860)
P. levigatus†† Rathbun, 1945
P. miocaenicus†† Müller, 1984a
P. miyaku†† Karasawa & Nobuhara, 2008
P. monspeliensis†† (A. Milne-Edwards, 1860)
P. neogenicus†† Müller, 1978
P. oblongus†† Rathbun, 1920a
P. obtusus†† A. Milne-Edwards, 1860
P. obvallatus†† Morris & Collins, 1991
P. pelagicus† (Linnaeus, 1758)
P. pirabaensis†† Martins-Neto, 2001
P. radobojanus†† (Bittner, 1884)
P. regulensis†† Van Straelen, 1939b
P. ristorii†† Karasawa, Schweitzer & Feldmann, 2008 (=*Neptunus convexus* Ristori, 1889a)
P. sanshianus†† Hu, 1984b
P. sebae (H. Milne Edwards, 1834 [in 1834-1840])
P. sijuensis†† (Satsangi & Parida, 1980)
P. spinimanus† Latreille, 1819
P. stenaspis†† (Bittner, 1884)
P. straeleni†† (Vía, 1948)
P. suessi†† (Bittner, 1875)
P. tenuis†† Rathbun, 1919 (claws only)
P. thalae†† Macarovici, 1970
P. tongfai†† Hu, 1981
P. vectensis†† (Carter, 1898)
P. viai†† Secretan *in* Philippe & Secretan, 1971

P. vicentinus†† (A. Milne-Edwards, 1860)
P. wynneanus†† (Stoliczka, 1871)
P. withersi†† (Glaessner, 1933b)
P. woodwardi†† Morris & Collins, 1991
P. yaucoensis†† Schweitzer, Iturralde-Vinent, Hetler & Velez-Juarbe, 2006
P. xantusii† (Stimpson, 1862b)
Pseudoachelous†† Portell & Collins, 2004 (0, 0, 1)
P. schindleri Portell & Collins, 2004 (type)
Rathbunella†† Collins *in* Collins et al., 2009 (0, 0, 1)
R. pentaspinosa†† Collins *in* Collins et al., 2009 (type)
Subfamily Thalamitinae† Paul'son, 1875
Charybdis† De Haan, 1833 (56, 8, 15)
C. acuta† (A. Milne-Edwards, 1869a)
C. annulata† (Fabricius, 1798)
C. antiqua†† (A. Milne-Edwards, 1860)
C. arabica†† (Woodward, 1905)
C. bimaculata† (Miers, 1886)
C. feriata† (Linnaeus, 1758)
C. fragilis†† (Müller, 1978)
C. fijiensis†† Rathbun, 1945
C. gigantica†† Hu & Tao, 1996
C. granulata† (De Haan, 1833)
C. hoffmeisteri†† Rathbun, 1945
C. japonica† (A. Milne-Edwards, 1861)
C. kilmeri†† Hu, 1984b
C. leei†† Hu & Tao, 1996
C. mathiasi†† Müller, 1984a
C. miles† (De Haan, 1835)
C. minuta†† Hu & Tao, 1996
C. monsoonis†† Hu & Tao, 1985
C. obtusa†† Hu & Tao, 1996
C. orientalis† Dana, 1852c
C. preferiata†† Hu & Tao, 1996
C. shihseihi†† Hu & Tao, 1979
C. sinhaleya†† Deraniyagala, 1958
Eocharybdis†† Beschin et al., 2002 (0, 0, 1)
E. cristata Beschin et al., 2002 (type)

Thalamita† Latreille, 1829 (87, 2, 3)
T. admete† (Herbst, 1803) (type)
T. crenata† (Latreille, 1829)
T. pleistocenica Hu & Tao, 1979
T. truncatata†† Hu & Tao, 1996
T. fani†† Hu & Tao, 1996
Thalamitoides† A. Milne-Edwards, 1869a (2, 1, 0)
T. quadridens† A. Milne-Edwards, 1869a (type)
Psammocarcinidae†† Beurlen, 1930a
Psammocarcinus†† A. Milne-Edwards, 1860 (0, 0, 1)
P. hericarti (Desmarest, 1822) (type)
Thiidae† Dana, 1852a
Thia† Leach, 1816 [imprint 1815] (1, 0, 1)
T. simplex†† Rathbun, 1945
Superfamily Trichodactyloidea H. Milne Edwards, 1853
Family **Trichodactylidae** H. Milne Edwards, 1853
Subfamily Dilocarcininae† Pretzmann, 1978
Sylviocarcinus† H. Milne Edwards, 1853 (4, 1, 0)
S. piriformis† (Pretzmann, 1968)
Subfamily Trichdactylinae H. Milne Edwards, 1853
No fossil representatives
Superfamily Bythograeoidea Williams, 1980
Family **Bythograeidae** Williams, 1980
No fossil representatives
Superfamily Carpilioidea Ortmann, 1893
Family **Carpiliidae**† Ortmann, 1893
Carpilius† Desmarest, 1822 (0, 3, 2)
C. convexus† (Forskål, 1775)
C. corallinus† (Herbst, 1783)
C. lwangi†† Hu & Tao, 2000
C. maculatus† (Linnaeus, 1758) (type)
C. petreus†† Beschin, Busulini, De Angeli & Tessier, 2007
Eocarpilius†† Blow & Manning, 1996 (0, 0, 4)
E. anomolus (Rathbun, 1935a)
E. antiquus (Glaessner, 1928)
E. blowi Feldmann, Bice et al., 1998
E. carolinensis Blow & Manning, 1996 (type)

Holcocarcinus†† Withers, 1924d (0, 0, 1)
H. sulcatus Withers, 1924d (type)
Ocalina†† Rathbun, 1929b (0, 0, 4)
O. floridana Rathbun, 1929b (type)
O. haldixoni Collins & Donovan, 2006
O. straeleni (Remy *in* Remy & Tessier, 1954)
O. sublevis Collins & Donovan, 2005
Palaeocarpilius†† A. Milne-Edwards, 1862c (0, 0, 14)
P. aquilinus Collins & Morris, 1973
P. almerae (Vía, 1941)
P. aquitanicus A. Milne-Edwards, 1862c
P. bispinosus Satsangi & Changkakoti, 1989
P. brodkorbi Lewis & Ross, 1965
P. ignotus A. Milne-Edwards, 1862c
P. intermedius Stubblefield, 1927
P. klipsteini (Von Meyer, 1862)
P. laevis Imaizumi, 1939
P. macrocheilus Desmarest, 1822 (type)
P. mississippiensis (Rathbun, 1935a)
P. rugifer Stoliczka, 1871
P. simplex Stoliczka, 1871 (= *P. anodon* Bittner, 1875)
P. valrovinensis (de Gregorio, 1895)
Paraocalina†† Beschin, Busulini, De Angeli & Tessier, 2007 (0, 0, 1)
P. multilobata Beschin et al., 2007 (type)
Proxicarpilius†† Collins & Morris, 1978 (0, 0, 1)
P. planifrons Collins & Morris, 1978 (type)

Palaeoxanthopsidae†† Schweitzer, 2003
Jakobsenius†† Schweitzer, 2005a (0, 0, 1)
J. cretaceous (Segerberg, 1900) (type)
Lobulata†† Schweitzer, Feldmann & Gingerich, 2004 (0, 0, 1)
L. lobulata (Feldmann, Casadío et al., 1995) (type)
Palaeoxantho†† Bishop, 1986a (0, 0, 1)
P. libertiensis Bishop, 1986a (type)
Palaeoxanthopsis†† Beurlen, 1958b (0, 0, 3)
P. cretacea (Rathbun, 1902a) (type)
P. meyapaquensis (Vega, Feldmann et al., 2001)

P. tylotus Schweitzer et al., 2008
Paraverrucoides†† Schweitzer, 2003 (0, 0, 1)
P. alabamensis (Rathbun, 1935a) (type)
Remia†† Schweitzer, 2003 (0, 0, 1)
R. africana (Remy *in* Remy & Tessier, 1954) (type)
Rocacarcinus†† Schweitzer, 2005a (0, 0, 1)
R. gerthi (Glaessner, 1930) (type)
Verrucoides†† Vega, Cosma et al., 2001 (0, 0, 2)
V. stenohedra Vega, Cosma et al., 2001
V. verrucoides (Collins & Rasmussen, 1992) (type)
Incertae sedis
Cancer verrucosus Schafhäutl, 1851
Tumidocarcinidae†† Schweitzer, 2005a
Baricarcinus†† Casadío et al., 2004 (0, 0, 1)
B. mariae Casadío et al., 2004 (type)
Cyclocorystes†† Bell, 1858 (0, 0, 3)
C. aldersoni Squires, 1980
C. pulchellus Bell, 1858 (type)
C. vulgatus (Quayle & Collins, 1981)
Lobonotus†† A. Milne-Edwards, 1863 (0, 0, 9)
L. australis Fritsch & Kafka, 1878
L. bakeri (Rathbun, 1935a)
L. brazosensis Stenzel, 1935
L. mexicanus Rathbun, 1930b
L. natchitochensis Stenzel, 1935
L. purdyi (Blow & Manning, 1996)
L. sandersi (Blow & Manning, 1998)
L. sculptus A. Milne-Edwards, 1863 (type)
L. sturgeoni (Feldmann, Bice et al., 1998)
Nitotacarcinus†† Schweitzer, Artal et al., 2007 (0, 0, 2)
N. bituberculatus (Collins & Jakobsen, 2003) (type)
N. canadensis Schweitzer, Feldmann, Ćosović, Ross & Waugh, 2009
Paratumidocarcinus†† Martins-Neto, 2001 (0, 0, 1)
P. marajoarus Martins-Neto, 2001 (type)
Pulalius†† Schweitzer, Feldmann, Tucker & Berglund, 2000 (0, 0, 2)
P. dunhamorum Schweitzer et al., 2000

P. vulgaris (Rathbun, 1926b) (type)
Titanocarcinus†† A. Milne-Edwards, 1863 (0, 0, 7)
T. briarti (Forir, 1887)
T. decor Schweitzer, Artal et al., 2007
T. faxeensis (Von Fischer-Benzon, 1866)
T. kambuhelensis Verhoff et al., 2009
T. raulinianus A. Milne-Edwards, 1863
T. serratifrons A. Milne-Edwards, 1863 (type)
T. subellipticus (Segerberg, 1900)
Tumidocarcinus†† Glaessner, 1960 (0, 0, 4)
T. dentatus Glaessner, 1960
T. giganteus Glaessner, 1960
T. tumidus (Woodward, 1876b) (type)
T. victoriensis (Crespin, 1947)
Xanthilites†† Bell, 1858 (0, 0, 2)
X. bowerbanki Bell, 1858 (type)
X. interpunctatus Schafhäutl, 1863
Zanthopsidae†† Vía, 1959
Fredericia†† Collins & Jakobsen, 2003 (0, 0, 1)
F. barsoei Collins & Jakobsen, 2003 (type)
Harpactocarcinus†† A. Milne-Edwards, 1862c (0, 0, 10)
H. dalmatius Schweitzer, Shirk et al., 2007
H. istriensis Bachmayer & Nosan, 1959
H. jacquoti A. Milne-Edwards, 1865c
H. macrodactylus (H. Milne Edwards *in* d'Archiac, 1850)
H. multidentatus Stubblefield, 1946
H. ovalis A. Milne-Edwards, 1862c
H. punctulatus A. Milne-Edwards, 1862c (type)
H. rotundatus A. Milne-Edwards, 1862c
H. wilkeningi (Bachmayer & Mundlos, 1968)
H. yozgatensis Schweitzer, Shirk et al., 2007
Harpactoxanthopsis†† Vía, 1959 (0, 0, 5)
H. bittneri (Lőrenthey, 1898a)
H. lutugini (Likharev, 1917)
H. quadrilobatus (Desmarest, 1922) (type)
H. souverbei (A. Milne-Edwards, 1862c)
H. villaltae Vía, 1959

Martinetta†† Blow & Manning, 1997 (0, 0, 1)
M. palmeri Blow & Manning, 1997 (type)
Neozanthopsis†† Schweitzer, 2003 (0, 0, 7)
N. achalzicus (Bittner, 1882)
N. americanus (Rathbun, 1928) (type)
N. bruckmanni (Von Meyer, 1862)
N. carolinensis (Rathbun, 1935a)
N. rathbunae (Stenzel, 1934)
N. sonthofenensis (Von Meyer, 1862)
N. tridentata (Von Meyer, 1862)
Zanthopsis†† McCoy, 1849 (0, 0, 12)
Z. bispinosa Bell, 1858
Z. dufourii (H. Milne Edwards *in* d'Archiac, 1850)
Z. errans Woods, 1922
Z. hendersonianus Rathbun, 1926b
Z. jacobi Van Straelen, 1925
Z. leachii (Desmarest, 1822) (type)
Z. milleri Roberts, 1955
Z. minor Stolley, 1890
Z. multispinosa Remy, 1960
Z. nodosa McCoy, 1849
Z. peytoni Stenzel, 1934
Z. unispinosa McCoy, 1849
Superfamily Pilumnoidoidea Guinot & Macpherson, 1987
Family **Pilumnoididae** Guinot & Macpherson, 1987
No fossil representatives
Superfamily Pilumnoidea Samouelle, 1819
Family **Galenidae**† Alcock, 1898
Subfamily Dentoxanthinae Števčić, 2005
No fossil representatives
Subfamily Galeninae† Alcock, 1898
Galene† De Haan, 1833 (0, 1, 3)
G. bispinosa† (Herbst, 1804) (= *Cancer trispinosa*†† Desmarest, 1822; = *Galene hainanensis*†† Hu & Tao, 1979; see Karasawa & Kato *in* Karasawa et al., 2008)
G. granulifera†† Lin, 1947
G. obscura†† A. Milne-Edwards, 1865c

G. stipata†† Morris & Collins, 1991
Subfamily Halimedinae† Alcock, 1898
Halimede† De Haan, 1833 (2, 2, 0)
H. fragifer† (De Haan, 1835)
H. octhodes† (Herbst, 1793)
Subfamily Parapanopinae Števčić, 2005
No fossil representatives
Pilumnidae† Samouelle, 1819
Subfamily Calmaniinae Števčić, 2005
No fossil representatives
Subfamily Eumedoninae† Dana, 1853
Santeella†† Blow & Manning, 1996 (0, 0, 1)
S. lillyae Blow & Manning, 1996 (type)
Viacarcinus†† Blow & Manning, 1996 (0, 0, 1)
V. druidi Blow & Manning, 1996 (type)
Subfamily Pilumninae† Samouelle, 1819
Actumnus† Dana, 1851b (26, 1, 2)
A. guinotae†† Karasawa & Inoue, 1992
A. squamosus† (De Haan, 1835)
A. telegdii†† (Müller, 1974a)
Budapanopeus†† Müller & Collins, 1991 (0, 0, 1)
B. denticulatus Müller & Collins, 1991 (type)
Eohalimede†† Blow & Manning, 1996 (0, 0, 3)
E. granosa Beschin et al., 2002
E. sandersi Blow & Manning, 1998
E. walleri Blow & Manning, 1996 (type)
Eopilumnus†† Beschin et al., 2002 (0, 0, 1)
E. checchii Beschin et al., 2002 (type)
Eumorphactaea†† Bittner, 1875 (0, 0, 2)
E. latifrons A. Milne-Edwards, 1880b
E. scissifrons Bittner, 1875 (type)
Galenopsis†† A. Milne-Edwards, 1865c (0, 0, 12)
G. americana Rathbun, 1935a
G. crassifrons A. Milne-Edwards, 1865c
G. depressa A. Milne-Edwards, 1872
G. fulangaensis Rathbun, 1945
G. gervilliana A. Milne-Edwards, 1865c
G. murchisoni A. Milne-Edwards, 1865c

G. pustulosa A. Milne-Edwards, 1865c

G. ristorii Checchia-Rispoli, 1907

G. schopeni Checchia-Rispoli, 1905

G. similis Bittner, 1875

G. typica A. Milne-Edwards, 1865c (type)

G. wulfi Förster & Mundlos, 1982

Glabropilumnus† Balss, 1932 (5, 1, 3)

G. fossatus†† Müller, 1996

G. kamiyai†† Karasawa, 1991

G. laevis† (Dana, 1852c)

G. soghensis†† De Angeli & Beschin, 2008

Heteropanope† Stimpson, 1859 [read 1858] (5, 0, 1)

H. wylliei†† Glaessner, 1933b

Lobogalenopsis†† Müller & Collins, 1991 (0, 0, 1)

L. quadrilobata (Lőrenthey, 1897 [German version 1898a]) (type)

Pilumnopeus† A. Milne-Edwards, 1867 (12, 0, 3)

P. dilatatus†† Müller, 1993

P. paratethyensis†† Müller, 1984a

P. tetenyensis†† Müller, 1984a

Pilumnus† Leach, 1816 [imprint 1815] (134, 5, 7)

P. cucaoensis†† Feldmann et al., 2005

P. fookimensis†† Collins et al., 2003

P. hirtellus† (Linnaeus, 1761)

P. hirtellus var. *villosa*†† Risso, 1826

P. mediterraneus†† (Lőrenthey, 1898a)

P. olivellai†† Müller, 1993

P. subequus†† Rathbun, 1919 (claws only)

P. sayi† Rathbun, 1897c

P. aff. *P. pannosus*† Rathbun, 1898b

P. aff. *P. spinossimus*† Rathbun, 1898b

P. villosissimus† (Rafinesque, 1814)

Subfamily Rhizopinae† Stimpson, 1858c

Arges† De Haan, 1835 (0, 0, 1)

A. parallelus†† De Haan, 1835

Cryptolutea† Ward, 1936 (5, 0, 2)

C. littoralis†† (Collins et al., 2003)

C. warreni†† Karasawa & Kato *in* Karasawa et al., 2008

Typhlocarcinus† Stimpson, 1858c (7, 0, 3)
T. galenoformis†† Hu & Tao, 1996
T. obtusus†† Karasawa, 1993
T. taiwanicus†† Hu, 1983
Subfamily Xenophthalmodinae Števčić, 2005
No fossil representatives
Incertae sedis
"*Titanocarcinus*" *kochi* Lőrenthey, 1898a
Superfamily Xanthoidea† MacLeay, 1838

Discussion. — Xanthoidea as treated herein differs from that of Ng et al. (2008) in including the Domeciidae and Trapeziidae following Karasawa & Schweitzer (2006).

Domeciidae† Ortmann, 1893
Jonesius† Sankarankutty, 1962 (1, 0, 2)
J. oligocenicus†† (Beschin, De Angeli & Checchi, 2001)
J. planus†† (Müller, 1996)
Palmyria† Galil & Takeda, 1986 (0, 1, 0)
P. palmyrensis† (Rathbun, 1923c)
Panopeidae† Ortmann, 1893
Subfamily Eucratopsinae† Stimpson, 1871a
Balcacarcinus†† Karasawa & Schweitzer, 2006 (0, 0, 1)
B. attenuatus (Bittner, 1875) (type)
Bittnereus†† Beschin, Busulini, De Angeli & Tessier, 2007 (0, 0, 1)
B. vicentinus (Bittner, 1875) (type)
Carinocarcinus†† Lőrenthey, 1898b (0, 0, 1)
C. zitteli Lőrenthey, 1898b (type)
Glyphithyreus†† Reuss, 1859 (= *Plagiolophus* Bell, 1858) (0, 0, 4)
G. ellipticus (Bittner, 1875)
G. formosus Reuss, 1859 (type)
G. sulcatus (Beurlen, 1939)
G. wetherelli Bell, 1858
Palaeograpsus†† Bittner, 1875 (0, 0, 1)
P. inflatus Bittner, 1875 (type)
Subfamily Panopeinae† Ortmann, 1893
Eurypanopeus† A. Milne-Edwards, 1880a (12, 1, 0)
E. depressus† (Smith, 1869b)

Eurytium† Stimpson, 1862a (3, 1, 2)
E. crenulatum†† Rathbun, 1918
E. granulosum†† Schweitzer, Velez-Juarbe et al., 2008
E. limosum† (Say, 1818)
Hexapanopeus† Rathbun, 1898b (10, 1, 0)
H. caribbaeus† (Stimpson, 1871a)
Laevicarcinus†† Lőrenthey *in* Lőrenthey & Beurlen, 1929 (0, 0, 3)
L. dockeryi Blow & Manning, 1996
L. egerensis Lőrenthey *in* Lőrenthey & Beurlen, 1929 (type)
L. lioyi Beschin, Busulini, De Angeli & Tessier, 2007
Lophopanopeus† Rathbun, 1898b (6, 3, 3)
L. corallinus†† Portell & Collins, 2004
L. diegensis† Rathbun, 1900
L. leucomanus† (Lockington, 1877a)
L. lockingtoni† Rathbun, 1900
L. maculoides†† Collins & Todd *in* Todd & Collins, 2005
L. toomeyorum†† Portell & Collins, 2004
Panopeus† H. Milne Edwards, 1834 [in 1834-1840] (15, 2, 20)
P. antepurpureus†† (Rathbun, 1918) (claw only)
P. baldwini†† (Kooser & Orr, 1973)
P. bessmanni†† Collins & Jakobsen, 2003
P. bolcensis†† Secretan, 1975
P. capanemaensis†† Martins Neto, 2001
P. estellensis†† Rathbun, 1935a (claw only)
P. granulineatus†† Müller & Collins, 1991
P. herbstii† H. Milne Edwards, 1834 [in 1824-1840]
P. incisus†† Beschin, Busulini, De Angeli & Tessier, 2007
P. jerseyensis†† Roberts, 1956 (claw only)
P. lauensis†† Rathbun, 1945
P. nanus†† Portell & Collins, 2004
P. olearis†† (Rathbun, 1926b)
P. piramidensis†† Casadío et al., 2005
P. prosakrolophos†† Schweitzer, 2000
P. rugosus† A. Milne-Edwards, 1880a

P. soledadensis†† (Rathbun, 1926b)
P. torus†† Schweitzer, 2000
P. tridentatus†† Rathbun, 1918 (claw only)
P. veintensis†† Vega et al., 2008
P. viai†† Müller, 1993
P. wronai†† Müller, 1984a
Tetraxanthus† Rathbun, 1898b (2, 1, 0)
T. rathbunae† (Chace, 1939)

Fossil **Panopeidae** not assigned to subfamily level
Pakicarcinus†† Schweitzer, Feldmann & Gingerich, 2004 (0, 0, 2)
P. orientalis (Collins & Morris, 1978) (type)
P. whittenensis (Glaessner, 1980)
Sereneopeus†† Collins, 2002 (0, 0, 2)
S. humilis Beschin, Busulini, De Angeli & Tessier, 2007
S. kempi (Quayle & Collins, 1981) (type)

Pseudorhombilidae† Alcock, 1900b
Nanoplax† Guinot, 1967 (0, 1, 0)
N. xanthiformis† (A. Milne-Edwards, 1880a)
Pseudorhombila† H. Milne Edwards, 1834 [in 1834-1840] (5, 0, 1)
P. patagonica†† Glaessner, 1933b

Tanaocheleidae Ng & Clark, 2000
No fossil representatives

Tetraliidae Castro, Ng & Ahyong, 2004
Tetralia† Dana, 1851e (10, 0, 1)
T. loerentheyi†† (Müller, 1975)

Trapeziidae† Miers, 1886
Archaeotetra†† Schweitzer, 2005b (0, 0, 1)
A. inornata Schweitzer, 2005b (type)
Eomaldivia†† Müller & Collins, 1991 (0, 0, 2)
E. pannonica Müller & Collins, 1991 (type)
E. trispinosa Müller & Collins, 1991
Paratetralia†† Beschin, Busulini, De Angeli & Tessier, 2007 (0, 0, 1)
P. convexa Beschin et al., 2007 (type)
Trapezia† Latreille, 1825 (22, 1, 3)
T. brevispinosa†† Karasawa, 1993

T. digitalis† Latreille, 1828

T. glaessneri†† Müller, 1975

T. prisca†† Portell & Collins, 2004

Xanthidae† MacLeay, 1838

Subfamily Actaeinae† Alcock, 1898

Actaea† De Haan, 1833 (23, 5, 5)

A. ? *acantha*† (H. Milne Edwards, 1834 [in 1834-1840])

A. ? *bifrons*† Rathbun, 1898b

A. calzadai†† Müller, 1984b

A. cretacea†† Rathbun, 1935a (probably not *Actaea*)

A. persica†† A. Milne-Edwards, 1865c

A. polyacantha† (Heller, 1861)

A. pura† Stimpson, 1859 [read 1858]

A. semblatae† Guinot, 1976

A. tropicana†† Hu & Tao, 1996

A. turcocampestris†† Müller, 1984a

Actaeodes† Dana, 1851b (4, 2, 0)

A. hirsutissima† (Ruppell, 1830)

A. tomentosus† (H. Milne Edwards, 1834 [in 1834-1840])

Actaeops†† Portell & Collins, 2004 (0, 0, 1)

A. frontalis Portell & Collins, 2004 (type)

Forestia† Guinot, 1976 (4, 0, 1)

F. tagauna†† Hu & Tao, 1996

Gaillardiellus† Guinot, 1976 (4, 1, 0)

G. orientalis† (Odhner, 1925)

Heteractaea† Lockington, 1877b (2, 1, 0)

H. lunata†† (A. Milne-Edwards & Lucas, 1843)

Paractaea† Guinot, 1969 (13, 1, 0)

P. philippinensis† (Ward, 1942)

Phlyctenodes†† A. Milne-Edwards, 1862c (0, 0, 8)

P. dalpiazi Fabiani, 1911

P. irregularis Ristori, 1896

P. krenneri Lőrenthey, 1897 [German version 1898a]

P. multituberculatus Beschin, Busulini, De Angeli & Tessier, 2007

P. nicolisi Bittner, 1884

P. pustulosus A. Milne-Edwards, 1862c

P. steinmanni Lőrenthey, 1902
P. tuberculosus A. Milne-Edwards, 1862c (type)
Pseudophlyctenodes†† Busulini et al., 2006 (0, 0, 1)
P. hantkeni (Lőrenthey, 1897 [German version 1898a]) (type)
Subfamily Antrocarcininae Ng & Chia, 1994
No fossil representatives
Subfamily Chlorodiellinae† Ng & Holthuis, 2007
Chlorodiella† Rathbun, 1897a (= *Chlorodius* H. Milne Edwards, 1834 [in 1834-1840]) (8, 3, 6)
C. cytherea† (Dana, 1852c)
C. juglans†† Müller, 1984a
C. junghuhni†† (Martin, 1880)
C. laevissima† (Dana, 1852c)
C. loczyi†† Müller, 1984a
C. mediterranea†† Lőrenthey *in* Lőrenthey & Beurlen, 1929
C. mediterranea tetenyensis†† Müller, 1984a
C. nigra† (Forskål, 1775)
C. occidentalis†† Portell & Collins, 2004
Cyclodius† Dana, 1851b (5, 3, 0)
C. granulosus† De Man, 1888
C. nitidus† (Dana, 1852c)
C. ungulatus† (H. Milne Edwards, 1834)
Pilodius† Dana, 1851b (14, 3, 3)
P. granulatus† Stimpson, 1859 [read 1858]
P. mediterraneus†† Lőrenthey, 1898a
P. nigrocrinitus† Stimpson, 1859 [read 1858]
P. parvus†† Karasawa, 1993
P. pugil† Dana, 1852c
P. vulgaris†† (Glaessner, 1928)
Subfamily Cymoinae† Alcock, 1898
Only unnamed species
Subfamily Etisinae† Ortmann, 1893
Etisus† H. Milne Edwards, 1834 [in 1834-1840] (21, 3, 3)
E. arduinoi†† Beschin, Busulini, De Angeli & Tessier, 2007
E. dentatus† (Herbst, 1785)

E. javanicus†† (Van Straelen, 1938)
E. laevimanus† Randall, 1839
E. pleistocenicus†† Hu & Tao, 2000
E. splendidus† Rathbun, 1906 (= *E. doucerei* Kowalski, 1923)

Subfamily Euxanthinae† Alcock, 1898
Euxanthus† Dana, 1851b (6, 0, 1)
E. multipunctulatus†† Hu & Tao, 1996
Glyptoxanthus† A. Milne-Edwards, 1879 [in 1873-1881] (8, 0, 1)
G. primitivus†† Müller, 1993
Hypocolpus† Rathbun, 1897a (11, 1, 0)
H. abbotti† (Rathbun, 1894)
Lipaesthesius† Rathbun, 1898a (1, 0, 1)
L. quadratus†† Rathbun, 1945
Medaeops† Guinot, 1967 (6, 0, 1)
M. megamiensis†† Karasawa, 1993
Medaeus† Dana, 1851b (4, 0, 3)
M. granulatus†† Martin, 1880
M. mbalavuensis†† Rathbun, 1945
M. ? *variabilis*†† Rathbun, 1945
Monodaeus† Guinot, 1967 (7, 1, 1)
M. bortolottii†† Delle Cave, 1988
M. couchii† (Couch, 1851)

Subfamily Kraussiinae Ng, 1993
No fossil representatives

Subfamily Liomerinae† Sakai, 1976
Liomera† Dana, 1851b (14, 6, 2)
L. cinctimana† (White, 1847c)
L. laevis† (A. Milne-Edwards, 1873c)
L. monticulosa† (A. Milne-Edwards, 1873c)
L. rubra† (A. Milne-Edwards, 1865d)
L. sublensis†† (Rathbun, 1945)
L. stimpsonii† (A. Milne-Edwards, 1965d)
L. transversa†† (Hu, 1983)
L. venosa† (H. Milne Edwards, 1834 [in 1834-1840])
Neoliomera† Odhner, 1925 (13, 4, 2)
N. insularis† (Adams & White, 1849)

N. intermedia† Odhner, 1925

N. kuohwai†† Hu, 1981

N. paleogenica†† Beschin, Busulini, De Angeli & Tessier, 2007

N. pubescens† (H. Milne Edwards, 1834 [in 1834-1840])

N. richteroides† Sakai, 1969

Neomeria†† Hu & Tao, 1996 (0, 0, 1)

N. lelingae (Hu, 1981) (type)

Subfamily Polydectinae† Dana, 1851b

Only unnamed species

Subfamily Speocarcininae† Števčić, 2005

Speocarcinus† Stimpson, 1862a (6, 0, 1)

S. berglundi†† Tucker et al., 1994

Subfamily Xanthinae† MacLeay, 1838

Cycloxanthops† Rathbun, 1897a (5, 1, 1)

C. eous†† Rathbun, 1945

C. novemdentatus† (Lockington, 1877a)

Demania† Laurie, 1906 (13, 4, 2)

D. caltripes† (Alcock, 1898)

D. chayiensis†† (Hu, 1981)

D. pilipinas†† Karasawa & Kato *in* Karasawa et al., 2008

D. reynaudii† (H. Milne Edwards, 1834 [in 1834-1840])

D. scaberrima† (Walker, 1887)

D. wardi† Garth & Ng, 1985

Garthiope† Guinot, 1990 (3, 1, 0)

G. spinipes† (A. Milne-Edwards, 1880a)

Lachnopodus† Stimpson, 1859 [read 1858] (4, 2, 1)

L. bidentatus† (A. Milne-Edwards, 1867)

L. murdjadjensis†† Saint Martin & Müller, 1988

L. subacutus† (Stimpson, 1859 [read 1858])

Leptodius† A. Milne-Edwards, 1863 (11, 0, 3)

L. crosnieri†† Karasawa, 1993

L. granulatus†† Portell & Collins, 2004

L. morrisi†† Karasawa, 1993

Liagore† De Haan, 1833 (2, 1, 0)

L. rubromaculata† (De Haan, 1835)

Macromedaeus† Ward, 1942 (5, 1, 0)
M. distinguendus† (De Haan, 1835)
Megamia†† Karasawa, 1993 (0, 0, 1)
M. anaglypta Karasawa, 1993 (type)
Metopoxantho† De Man, 1904 (0, 1, 0)
M. martini† De Man, 1904
Micropanope† Stimpson, 1871a (7 confirmed and 4 questionable, 2, 1)
M. ? nuttingi† (Rathbun, 1898b)
M. pulcherrima†† Portell & Collins, 2004
M. ? truncatifrons† Rathbun, 1898b
Nanocassiope Guinot, 1967 (5, 1, 0)
N. alcocki† (Rathbun, 1902a)
Palaeoxanthops†† Karasawa, 1993 (0, 0, 2)
P. minutus Karasawa, 1993 (type)
P. okumurai Karasawa, 1993
Paraxanthias† Odhner, 1925 (5, 0, 1)
P. fujiyamai†† Karasawa, 1993
Xanthias† Rathbun, 1897a (12, 1, 2)
X. maculatus† (Forskål, 1775)
X. miocenicus†† Collins et al., 2003
X. similis†† Hu, 1983
Xantho† Leach, 1816 [imprint 1815] (4, 1, 4)
X. agassizi†† Robineau-Desvoidy, 1849
X. hydrophilus† (Herbst, 1790) (=*X. floridus* Montagu, 1813; *X. incisus* Leach, 1814)
X. lovisatoi†† (Lőrenthey, 1907a [German version 1909a])
X. moldavicus†† (Yanakevich, 1977)
X. vitiensis†† (Rathbun, 1945)
Xanthodius† Stimpson, 1862a (7, 0, 1)
X. kambaraensis†† Rathbun, 1945
Subfamily Zalasiinae† Serène, 1968
Banareia† A. Milne-Edwards, 1869b (14, 2, 0)
B. kraussi† (Heller, 1861)
B. odhneri† Sakai, 1974
Subfamily Zosiminae† Alcock, 1898
Atergatis† De Haan, 1833 (11, 3, 3)

A. dilatatus† De Haan, 1835
A. dubius†† A. Milne-Edwards, 1862c
A. granulatus† (De Man, 1889) (=*A. kymikis* Hu & Tao, 1996)
A. laddi†† Rathbun, 1945
A. latissimus† (H. Milne Edwards, 1834 [in 1834-1840])
A. pyroticus†† Hu & Tao, 1996
Platypodia† Bell, 1836 (9, 1, 0)
P. granulosa† (Rüppell, 1830)
Serenius† Guinot, 1976 (5, 1, 0)
S. ceylonicus† (Laurie, 1906)
Zosimus† Leach *in* Desmarest, 1825 (5, 1, 2)
Z. aeneus† (Linnaeus, 1758)
Z. jeneweni†† Hu & Tao, 2000
Z. poxus†† Hu & Tao, 1996
Fossil Xanthidae not assigned to subfamily
Haydnella†† Müller, 1984a (0, 0, 4)
H. maladensis Beschin, Busulini, De Angeli & Tessier, 2007
H. oligocenica De Angeli & Beschin, 2008
H. pulchellus (A. Milne-Edwards, 1864)
H. steiningeri Müller, 1984a (type)
Nogarolia†† Beschin, Busulini, De Angeli & Tessier, 1994 (0, 0, 1)
N. mirabilis Beschin et al., 1994 (type)
Sculptoplax†† Müller & Collins, 1991 (0, 0, 1)
S. rigida Müller & Collins, 1991 (type)
Superfamily Xanthoidea: incertae sedis
Actaeites†† Müller & Collins, 1991 (0, 0, 1)
A. lobatus Müller & Collins, 1991 (type)
Actaeopsis†† Carter, 1898 (0, 0, 1)
A. wiltshirei Carter, 1898 (type)
Caprocancer†† Müller & Collins, 1991 (0, 0, 1)
C. altus Müller & Collins, 1991 (type)
Colpocaris†† Von Meyer, 1862 (0, 0, 1)
C. bullatus (Von Meyer, 1862) (type)
Cretichlorodius†† Fraaye, 1996b (0, 0, 1)
C. enciensis Fraaye, 1996b (type)

Eoxanthias†† Hu & Tao, 1996 (0, 0, 1)
E. keelonicus Hu & Tao, 1996 (type)
Megaxantho†† Vega, Feldmann et al., 2001 (0, 0, 1)
M. zoque Vega, Feldmann et al., 2001 (type)
Muelleroplax†† ***nomen novum*** (= *Eoplax*†† Müller & Collins, 1991) (0, 0, 2)
M. minima (Müller & Collins, 1991) (type)
M. miniscula (Beschin, Busulini, De Angeli & Tessier, 2007)
Paraxanthosia†† Müller & Collins, 1991 (0, 0, 2)
P. budensis Müller & Collins, 1991 (type)
P. tuberculata Beschin et al., 2005
Pilumnomimus†† Müller & Collins, 1991 (0, 0, 1)
P. planidentatus Müller & Collins, 1991 (type)
Pregeryona†† Hu & Tao, 1996 (0, 0, 1)
P. taiwanica Hu & Tao, 1996 (type)
Prochlorodius†† Müller & Collins, 1991 (0, 0, 1)
P. ellipticus Müller & Collins, 1991 (type)
Styrioplax†† Glaessner, 1969 (= *Microplax* Glaessner, 1928) (0, 0, 1)
S. exiguus (Glaessner, 1928) (type)
Syphax†† A. Milne-Edwards, 1864 (0, 0, 1)
S. crassus A. Milne-Edwards, 1864 (type)
Thelecarcinus†† Böhm, 1922 (0, 0, 1)
T. guembeli Böhm, 1922 (type)
Thelphusograpsus Lőrenthey, 1902 (0, 0, 1)
T. laevis Lőrenthey, 1902 (type)
"*Titanocarcinus*" *euglyphos* Bittner, 1875
"*Titanocarcinus*" *mammillatus* Secretan, 1964
"*Titanocarcinus*" *polonicus* Fraaye, 1994
"*Titanocarcinus*" *sismondae* A. Milne-Edwards, 1864
"*Titanocarcinus*" *zoellneri* Bachmayer & Mundlos, 1968
"*Tumidocarcinus*" *foersteri* Aguirre-Urreta et al., 1995
Wanga†† Hu & Tao, 1996 (0, 0, 1)
W. minuta Hu & Tao, 1996 (type)
Woodbinax†† Stenzel, 1952 (0, 0, 1)
W. texanus Stenzel, 1952 (type)
"*Xanthilites*" *traubi* Förster, 1970b

"*Xantho*" *edwardsi* Sismonda, 1846
"*Xanthosia*" [*sensu lato*] (0, 0, 13)
X. americana Rathbun, 1930b (type appears to be missing)
X. buteonis Wright & Collins, 1972
X. danielae Collins & Breton, 2009
X. elegans Roberts, 1962
X. fischeri (A. Milne-Edwards, 1862c)
X. gracilis Jakobsen & Collins, 1997
X. occidentalis Bishop, 1991
X. robertsi Secretan, 1982
X. semiornata Jagt et al., 1991
X. socialis Van Bakel, Fraaije & Jagt, 2005
X. spinosa Bishop, 1991
X. zoquiapensis Fraaije, Vega, Van Bakel & Garibay-Romero, 2006
Xanthosioides†† Collins & Breton, 2009 (0, 0, 1)
X. delicata (Fraaye, 1996c) (type)
Superfamily Eriphioidea† MacLeay, 1838

Discussion. — We follow Karasawa & Schweitzer (2006) in defining the Eriphioidea and its constituent families.

Eriphiidae† MacLeay, 1838
Eriphia† Latreille, 1817b (6, 1, 2)
E. cocchii†† Ristori, 1886
E. gonagra xaymacaensis†† Collins & Donovan, 1997
E. verrucosa† (Forskål, 1775) [= *E. spinifrons* (Herbst, 1782)]
Hypothalassiidae† Karasawa & Schweitzer, 2006
Lathahypossia†† Schweitzer, Artal et al., 2007 (0, 0, 1)
L. aculeata (Busulini et al., 1984) (type)
Oziidae† Dana, 1851b
Epixanthus† Heller, 1862 (4, 1, 0)
E. frontalis† (H. Milne Edwards, 1834 [in 1834-1840])
Lydia† Gistl, 1848 (3, 0, 1)
L. kianobesana†† Hu, 1983
Menippe† De Haan, 1833 (4, 2, 8)
M. brevimanus†† Rathbun, 1945
M. burnsi†† Rathbun, 1935a

M. cretacea†† Rathbun, 1935a (probably not *Menippe*)
M. floridana†† Rathbun, 1935a
M. frescoensis†† Remy, 1960
M. mercenaria† (Say, 1818)
M. nodifrons† Stimpson, 1862a
M. oceanica†† Rathbun, 1945
M. jakesonensis†† Rathbun, 1935a
M. anomala†† Rathbun, 1935a
Ozius† H. Milne Edwards, 1834 [in 1834-1840] (13, 0, 1)
O. collinsi†† Karasawa, 1992b
Pseudocarcinus† H. Milne Edwards, 1834 [in 1834-1840] (0, 1, 0)
P. gigas† (Lamarck, 1818)
Platyxanthidae† Guinot, 1977
Pelaeus† Eydoux & Souleyet, 1842 (1, 0, 1)
P. chauvinii†† (de Berville, 1857)
Pseudoziidae† Alcock, 1898
Archaeozius†† Schweitzer, 2003 (0, 0, 1)
A. occidentalis (Schweitzer et al., 2000) (type)
Euryozius† Miers, 1886 (6, 0, 2)
E. angustus†† Karasawa, 1993
E. bidentatus†† Karasawa, 1993
Priabonocarcinus†† Müller & Collins, 1991 (0, 0, 1)
P. gallicus Müller & Collins, 1991 (type)
Santeexanthus†† Blow & Manning, 1996 (0, 0, 1)
S. wardi Blow & Manning, 1996 (type)
Tongapapaka†† Feldmann, Schweitzer & McLaughlin, 2006 (0, 0, 1)
T. motunauensis Feldmann et al., 2006 (type)
Superfamily Goneplacoidea† MacLeay, 1838
Family **Acidopsidae** Števčić, 2005
No fossil representatives
Carinocarcinoididae†† Karasawa & Kato, 2003
Bicarinocarcinus†† Glaessner & Secretan, 1987 (0, 0, 2)
B. collinsi Glaessner & Secretan, 1987 (type)
B. tumidus Collins et al., 2003
Carinocarcinoides†† Karasawa & Fudouji, 2000 (0, 0, 2)
C. angustifrons (Karasawa, 1993) (type)

C. carinatus Karasawa & Fudouji, 2000
Chasmocarcinidae† Serène, 1964
Subfamily Chasmocarcininae† Serène, 1964
Chasmocarcinus† Rathbun, 1898b (15, 0, 3)
C. guerini†† (Vía, 1959)
C. robertsi†† Blow & Bailey, 1992
C. seymourensis†† Feldmann & Zinsmeister, 1984b
Collinsius†† Karasawa, 1993 (0, 0, 1)
C. simplex Karasawa, 1993 (type)
Falconoplax†† Van Straelen, 1933b (0, 0, 2)
F. bicarinella Collins & Morris, 1976
F. kugleri Van Straelen, 1933b (type)
Gillcarcinus†† Collins & Morris, 1978 (0, 0, 1)
G. amphora Collins & Morris, 1978 (type)
Mioplax†† Bittner, 1884 (0, 0, 1)
M. socialis Bittner, 1884 (type)
Orthakrolophos†† Schweitzer & Feldmann, 2001b (0, 0, 3)
O. bartonensis (Quayle & Collins, 1981)
O. bittneri (Morris & Collins, 1991)
O. depressus (Quayle & Collins, 1981) (type)
Subfamily Megaesthesiinae Števčić, 2005
No fossil representatives
Subfamily Trogloplacinae Guinot, 1986
No fossil representatives
Conleyidae Števčić, 2005
No fossil representatives
Euryplacidae† Stimpson, 1871a
Chirinocarcinus†† Karasawa & Schweitzer, 2004 (0, 0, 1)
C. wichmanni (Feldmann, Casadío et al., 1995) (type)
Chlinocephalus†† Ristori, 1886 (0, 0, 2)
C. demissifrons Ristori, 1886 (type) (= *Titanocarcinus sculptus* Ristori, 1891a)
C. subovalis (Ristori, 1886)
Corallicarcinus†† Müller & Collins, 1991 (0, 0, 2)
C. planus Müller & Collins, 1991
C. spinosus (Lőrenthey *in* Lőrenthey & Beurlen, 1929) (type)
Eucrate† De Haan, 1835 (11, 1, 2)

E. crenata† De Haan, 1835

E. martini†† Rathbun, 1926b

E. puliensis†† Hu & Tao, 1996

Euryplax† Stimpson, 1862a (3, 0, 1)

E. culebrensis†† Rathbun, 1918

Orbitoplax†† Tucker & Feldmann, 1990 (0, 0, 3)

O. plafkeri Tucker & Feldmann, 1990 (type)

O. tuckerae Schweitzer, 2000

O. weaveri (Rathbun, 1926b)

Palaeopsopheticus†† Hu & Tao, 1996 (0, 0, 1)

P. shujenae Hu & Tao, 1996 (type)

Simonellia†† Vinassa de Regny, 1897 (0, 0, 1)

S. quircensis Vinassa de Regny, 1897 (type)

Stoaplax†† Vega, Cosma et al., 2001 (0, 0, 1)

S. nandachare Vega, Cosma et al., 2001 (type)

Viaplax†† Karasawa & Kato, 2003 (0, 0, 1)

V. urpiniana (Vía, 1959) (type)

Goneplacidae† MacLeay, 1838

Amydrocarcinus†† Schweitzer, Feldmann, González-Barba & Vega, 2002 (0, 0, 1)

A. dantei Schweitzer et al., 2002 (type)

Carcinoplax† H. Milne Edwards, 1852 *sensu lato* (19, 2, 12)

C. antiqua†† (Ristori, 1889b) (= *C. senecta* Imaizumi, 1961)

C. granulimana†† Karasawa & Inoue, 1992

C. imperfecta†† Karasawa & Inoue, 1992

C. longimana† (De Haan, 1833)

C. mongosungi†† Hu & Tao, 1985

C. prepurpurea†† Hu & Tao, 2000

C. prisca†† Imaizumi, 1961 (= *Pilumnoplax petrificus* Hu & Tao, 1996)

C. proavita†† (Glaessner, 1960)

C. purpurea† Rathbun, 1914 (= *C. linae* Hu & Tao, 1996)

C. shukumi†† Hu & Tao, 1985

C. temikoensis†† Feldmann & Maxwell, 1990

C. thongi†† Hu & Tao, 1985

C. tsengi†† Hu & Tao, 1996
C. wuhsini†† Hu & Tao, 1999
Goneplax† Leach, 1814 (3, 1, 5)
G. arenicola†† (Glaessner, 1960)
G. ? *carnuntinus*†† (Bachmayer, 1953 [imprint 1951])
G. craverii†† Crema, 1895
G. formosa†† Ristori, 1886 (= *G. meneghinii* Ristori, 1886)
G. rhomboides† (Linnaeus, 1758)
G. saccoi†† Crema, 1895 (= *G. gulderi* Bachmayer, 1953b)
Icriocarcinus†† Bishop, 1988c (0, 0, 1)
I. xestos Bishop, 1988c (type)
Kowaicarcinus†† Feldmann, Schweitzer, Maxwell & Kelley, 2008 (0, 0, 1)
K. maxwellae Feldmann et al., 2008 (type)
Magyarcarcinus†† Schweitzer & Karasawa, 2004 (0, 0, 1)
M. loczyanus (Lőrenthey, 1898a) (type)
Ommatocarcinus† White, 1852 (3, 1, 4)
O. arenicola†† Glaessner, 1960
O. corioensis†† (Creswell, 1886)
O. macgillivrayi† White, 1852
O. sankuei†† Hu & Tao, 2004
O. taiwanicus†† Hu & Tao, 1996
Psopheticus† Wood-Mason, 1892 (3, 1, 1)
P. shujenae†† (Hu & Tao, 1996)
P. stridulans† Wood-Mason, 1892
Hexapodidae† Miers, 1886
Goniocypoda†† Woodward, 1867 (0, 0, 10)
G. collinsi Crane & Quayle, 1986
G. edwardsi Woodward, 1867 (type)
G. elmorensis Crane & Quayle, 1986
G. quaylei Crane, 1981
G. rajasthanica Glaessner & Rao, 1960
G. sindensis Glaessner, 1933b
G. sulcata Carter, 1898
G. transsilvanica Bittner, 1893
G. tessieri Remy *in* Remy & Tessier, 1954

- - *G. verheyeni* Fraaije, Van Bakel & Jagt, 2007
  - *Hexapus*† De Haan, 1833 (4, 1, 4)
    - *H. anfractus*† (Rathbun, 1909)
    - *H. decapoda*†† (Morris & Collins, 1991)
    - *H. granuliformis*†† Karasawa & Kato *in* Karasawa et al., 2008
    - *H. nakajimai*†† Imaizumi, 1959b
    - *H. pinfoldi*†† Collins & Morris, 1978
  - *Palaeopinnixa*†† Vía, 1966a (0, 0, 10)
    - *P. eocenica* (Woods, 1922)
    - *P. granulosa* Schweitzer & Feldmann, 2002
    - *P. heckeri* (Birshtein, 1956)
    - *P. intermedia* (Collins & Morris, 1976)
    - *P. mytilicola* Vía, 1966 (type)
    - *P. perornata* Collins & Morris, 1976
    - *P. prima* (Rathbun, 1918)
    - *P. rathbunae* Schweitzer et al., 2000
    - *P. rocaensis* (Feldmann, Casadío et al., 1995)
    - *P. rotunda* Schweitzer & Feldmann, 2001b
  - *Stevea*† Manning & Holthuis, 1981 (1, 0, 1)
    - *S. cesarii*†† Beschin et al., 1994
- **Litocheiridae**† Števčić, 2005
  - *Gollincarcinus*†† Beschin & De Angeli, 2004 (0, 0, 1)
    - *G. levis* Beschin & De Angeli, 2004 (type)
  - *Lessinioplax*†† Beschin & De Angeli, 2004 (0, 0, 2)
    - *L. rugosa* Beschin & De Angeli, 2004
    - *L. simplex* Beschin & De Angeli, 2004 (type)
  - *Maingrapsus*†† Tessier et al., 1999 (0, 0, 1)
    - *M. quadratus* Tessier et al., 1999 (type)
  - *Paracorallicarcinus*†† Tessier et al., 1999 (0, 0, 2)
    - *P. arcanus* Tessier et al., 1999 (type)
    - *P. tricarinatus* Schweitzer, Feldmann, Gonzáles-Barba & Ćosović, 2006
- **Martinocarcinidae**†† Schweitzer, Feldmann & Bonadio, 2009
  - *Martinocarcinus*†† Böhm, 1922 (0, 0, 1)
    - *M. ickeae* (Böhm, 1922) (type)
- **Mathildellidae**† Karasawa & Kato, 2003
  - *Branchioplax*†† Rathbun, 1916 (0, 0, 10)

B. albertii De Angeli & Beschin, 2002
B. ballingi Remy *in* Remy & Tessier, 1954
B. ? *bidentata* Birshtein, 1956
B. carmanahensis (Rathbun, 1926b)
B. concinna Quayle & Collins, 1981
B. parva Beschin, Busulini, De Angeli & Tessier, 2007
B. pentagonalis (Yokoyama, 1911)
B. rossii De Angeli & Beschin, 2008
B. sulcata Müller & Collins, 1991
B. washingtoniana Rathbun, 1916 (type) (=*Neopilumnoplax hannibalanus* Rathbun, 1926b)

Coeloma†† A. Milne-Edwards, 1865c (only confirmed species listed) (0, 0, 6)
C. balticum Schlüter, 1879
C. granulosum A. Milne-Edwards, 1881b
C. latifrons Förster & Mundlos, 1982
C. macoveii Lăzărescu, 1959
C. taunicum Von Meyer, 1862
C. vigil A. Milne-Edwards, 1865c (type)

Coeloma [*sensu lato*] (incl. *Litoricola* Woodward, 1873) (0, 0, 13)
C. bicarinatum Ravn, 1903
C. birsteini Ilyin, 2005
C. credneri Noetling, 1881
C. dentatum Woodward, 1873
C. egerense Lőrenthey *in* Lőrenthey & Beurlen, 1929
C. glabra (Woodward, 1873) (type of *Litoricola*)
C. incarinatum Sorgenfrei, 1940
C. isseli Ristori, 1886
C. macrodactylus (Van Straelen, 1925)
C. (*Litoricola*) *martinezensis* Rathbun, 1926b
C. rupeliense Stainier, 1887
C. sanatium Ristori, 1886
C. vareolata Lőrenthey, 1898b

Tehuacana†† Stenzel, 1944b (0, 0, 3)
T. tehuacana Stenzel, 1944b (type)
T. schweitzerae Vega et al., 2008
T. ? *meridionalis* (Secretan, 1961)

Progeryonidae Števčić, 2005
No fossil representatives
Scalopidiidae Števčić, 2005
No fossil representatives
Vultocinidae Ng & Manuel-Santos, 2007
No fossil representatives
Superfamily Bellioidea Dana, 1852a
Family **Belliidae**† Dana, 1852a
Subfamily Belliinae Dana, 1852a
No recorded fossils
Subfamily Heteroziinae Števčić, 2005
No recorded fossils
Superfamily Potamoidea Ortmann, 1896
Potamidae† Ortmann, 1896
Subfamily Potaminae† Ortmann, 1896
Potamon† Savigny, 1816 (17, 0, 6)
P. antiquum†† Szombathy, 1916
P. castellinense†† (Szombathy, 1916)
P. ibericum†† Bachmayer & Pretzmann, 1971
P. quenstedti†† (Zittel, 1885)
P. sivalense†† Glaessner, 1933b
P. speciosum†† (Von Meyer, 1862)
Subfamily Potamiscinae† Bott, 1970
Geothelphusa† Stimpson, 1858c (48, 2, 0)
G. dehaani† (White, 1847b)
G. tenuimanus† (Miyake & Minei, 1965)
Potamonautidae† Bott, 1970
Subfamily Hydrothelphusinae Bott, 1955
No fossil representatives
Subfamily Potamonautinae† Bott, 1970
Potamonautes† MacLeay, 1838 (70, 1, 1)
P. niloticus† (H. Milne Edwards, 1837 [in 1834-1840])
P. tugenensis†† Morris, 1976
Tanzanonautes†† Feldmann et al., 2007 (0, 0, 1)
T. tuerkai Feldmann et al., 2007 (type)
Superfamily Pseudothelphusoidea Ortmann, 1893
Family **Pseudothelphusidae**† Ortmann, 1893
Subfamily Epilobocerinae Smalley, 1964

No fossil representatives
Subfamily Pseudothelphusinae† Ortmann, 1893
Questionable fossil record
Superfamily Gecarcinucoidea Rathbun, 1904b
Family **Gecarcinucidae** Rathbun, 1904b
No fossil representatives
Superfamily Cryptochiroidea Paul'son, 1875
Family **Cryptochiridae** Paul'son, 1875
No fossil representatives
Subsection Thoracotremata Guinot, 1977
Superfamily Pinnotheroidea De Haan, 1833
Family **Pinnotheridae**† De Haan, 1833
Subfamily Pinnothereliinae† Alcock, 1900b
Pinnixa† White, 1846 (54, 1, 7)
P. aequipunctata†† Morris & Collins, 1991
P. faba† (Dana, 1851f)
P. galliheri†† Rathbun, 1932c
P. microgranulosa†† Collins et al., 2003
P. miocenica†† Rathbun, 1932c
P. montereyensis†† Rathbun, 1932c
P. navidadensis†† Feldmann et al., 2005
P. omega†† Morris & Collins, 1991
Subfamily Pinnotherinae† De Haan, 1833
Pinnotheres† Bosc, 1802 (72, 0, 3?)
P. ? *araucana*†† Philippi, 1887
P. ? *elatus*†† A. Milne-Edwards, 1873a
P. ? *promaucanaus*†† Philippi, 1887
Viapinnixa†† Schweitzer & Feldmann, 2001b (0, 0, 3)
V. alvarezi Vega, Cosma et al., 2001
V. nodosa (Collins & Rasmussen, 1992) (type)
V. perrilliatae Vega, Nyborg et al., 2007
Zaops† Rathbun, 1900 (1, 1, 0)
Z. ostreus† (Say, 1817)
Incertae sedis
Propinnotheroides†† Bishop & Palmer, 2006
P. orangeburgensis†† Bishop & Palmer, 2006 (type)
Superfamily Ocypodoidea Rafinesque, 1815
Family **Camptandriidae**† Stimpson, 1858c

Paracleistostoma† De Man, 1895 (8, 0, 1)
P. miocaenica†† Müller, 1998
Dotillidae Stimpson, 1858c
No fossil representatives
Heloeciidae H. Milne Edwards, 1852
No fossil representatives
Macrophthalmidae† Dana, 1851f
Subfamily Ilyograpsinae Števčić, 2005
No fossil representatives
Subfamily Macrophthalminae† Dana, 1851f
Macrophthalmus† Latreille *in* Desmarest, 1823 (= *Cyphoplax*†† Haime, 1855; *Palaeoplax*†† A. Milne-Edwards & Brocchi, 1879) (subgenera not identified) (45, 7, 3)
M. aquensis†† A. Milne-Edwards & Brocchi, 1879 (= *M. vindobonensis*†† Glaessner, 1924)
M. depressus† Rüppell, 1830
M. japonicus† (De Haan, 1835)
M. latifrons† Haswell, 1882
M. latreillei† (Desmarest, 1822) (= *Cyphoplax impressa* Desmarest, 1822; *C. incerta*†† Desmarest, 1822; *C. incisus*†† Desmarest, 1822; *M. granulosus*† De Man, 1904; *M. officinalis*†† Hu & Tao, 1979)
M. leptophthalmus† (H. Milne Edwards, 1852) (= *M. taiwanicus*†† Hu & Tao, 1996)
M. serratus† Adams & White, 1848
M. setosus† H. Milne Edwards, 1852
M. viai†† Karasawa & Inoue, 1992
Subgenera identified:
M. (*Hemiplax*) Heller, 1865 (0, 1, 1)
M. (*H.*) *hirtipes*† Heller, 1865
M. (*H.*) *major*†† Glaessner, 1960
M. (*Mareotis*) Barnes, 1967 (1 fossil)
M. (*M.*) *wilfordi*†† Morris & Collins, 1991
Subfamily Tritodynamiinae† Števčić, 2005
Tritodynamia† Ortmann, 1894 (7, 0, 1)
T. miocaenica†† Müller, 2006
Mictyridae Dana, 1851f
No fossil representatives

Ocypodidae† Rafinesque, 1815
Subfamily Ocypodinae† Rafinesque, 1815
Ocypode† Weber, 1795 (23, 3, 1)
O. cf. *O. africana*† De Man, 1881
O. cursor† (Linnaeus, 1758)
O. quadrata† (Fabricius, 1787) [= *O. albicans*† Bosc, 1802]
O. vericoncava†† Casadío et al., 2005
Subfamily Ucinae† Dana, 1851f
Uca† Leach, 1814 (92, 4, 6)
U. antiqua†† Brito, 1972
U. galapagensis† Rathbun, 1902b [= *Gelasimus macrodactylus* H. Milne Edwards & Lucas, 1843]
U. hamlini†† Rathbun, 1926b
U. inaciobritoi†† Martins-Neto, 2001
U. maracoani† (Latreille, 1802)
U. miocenica†† Artal, 2007
U. nitida†† (Desmarest, 1822)
U. oldroydi†† Rathbun, 1926b
U. pugnax† (Smith, 1870)
U. subcylindrica† (Stimpson, 1862a)
Ucididae Števčić, 2005
No fossil representatives
Xenophthalmidae† Stimpson, 1858c
Subfamily Anomalifrontinae Rathbun, 1929c
No fossil representatives
Subfamily Xenophthaminae† Stimpson, 1858c
Xenophthalmus† White, 1846 (2, 0, 1)
X. subitus†† Morris & Collins, 1991
Superfamily Palicoidea† Bouvier, 1898
Family **Crossotonotidae**† Moosa & Serène, 1981
Crossotonotus† A. Milne-Edwards, 1873c (4, 0, 1)
C. diosdensis†† Müller, 1984a
Palicidae† Bouvier, 1898
Eopalicus†† Beschin, Busulini, De Angeli & Tessier, 1996b (0, 0, 3)
E. imbricatus De Angeli & Beschin, 2000
E. semicarinatus De Angeli & Beschin, 2000

E. squamosus Beschin et al., 1996b (type)
Palicus† Philippi, 1838 (23, 0, 1)
P. hungaricus†† Müller, 2006
Spinipalicus†† Beschin & De Angeli, 2003 (0, 0, 1)
S. italicus Beschin & De Angeli, 2003 (type)
Superfamily Grapsoidea MacLeay, 1838
Family **Gecarcinidae**† MacLeay, 1838
Cardisoma† Latreille, 1825 (3, 1, 1)
C. guanhumi† Latreille, 1828
C. planum†† Rathbun, 1945
Glyptograpsidae Schubart, Cuesta & Felder, 2002
No fossil representatives
Grapsidae† MacLeay, 1838
Litograpsus†† Schweitzer & Karasawa, 2004 (0, 0, 1)
L. parvus (Müller & Collins, 1991) (type)
Metopograpsus† H. Milne Edwards, 1853 (6, 0, 1)
M. badenis†† Müller, 2006
Miograpsus†† Fleming, 1981 (0, 0, 1)
M. papaka Fleming, 1981 (type)
Pachygrapsus† Randall, 1839 (14, 0, 1)
P. hungaricus†† Müller, 1974b
Planes† Bowditch, 1825 (3, 0, 1)
P. prior†† (Smirnov, 1929)
Plagusiidae Dana, 1851f
No fossil representatives
Sesarmidae† Dana, 1851f
Neosarmatium† Serène & Soh, 1970 (15, 1, 0)
N. smithi† (H. Milne Edwards, 1853)
Sesarma† Say, 1817 (17, 0, 2)
S. paraensis†† Beurlen, 1958a
S. primigenium†† Collins et al., 2009
Varunidae† H. Milne Edwards, 1853
Subfamily Asthenognathinae† Stimpson, 1858c
Asthenognathus† Stimpson, 1858c (4, 0, 6)
A. cornishorum†† Schweitzer & Feldmann, 1999
A. globosa†† (Karasawa, 1990)
A. laverdensis†† De Angeli & Garassino, 2006b
A. microspinus†† Casadío et al., 2004

A. rakosensis†† Müller, 2006
A. urretae†† Schweitzer & Feldmann, 2001b
Globihexapus†† Schweitzer & Feldmann, 2001b (0, 0, 2)
G. kosekii Kato, 2005
G. paxillus Schweitzer & Feldmann, 2001b (type)
Subfamily Cyclograpsinae H. Milne Edwards, 1853
Austrohelice† Sakai, Türkay & Yang, 2001 (1, 0, 1)
A. manneringi†† Feldmann, Schweitzer, Maxwell & Kelley, 2008
Cyclograpsus† H. Milne Edwards, 1837 [in 1834-1840] (18, 0, 3)
C. directus†† Karasawa, 1989
C. lwangi†† Hu & Tao, 1999
C. rectangularis†† Karasawa, 1989
Miosesarma†† Karasawa, 1989 (0, 0, 2)
M. japonicum Karasawa, 1989 (type)
M. naguraense Kato, 1996
Subfamily Gaeticinae Davie & Ng, 2007
Sestrostoma† Davie & Ng, 2007 (3, 0, 1)
S. hichiro†† (Karasawa & Tanaka, 2005)
Subfamily Thalassograpsinae Davie & Ng, 2007
No fossil representatives
Subfamily Varuninae† H. Milne Edwards, 1853
Brachynotus† De Haan, 1833 (4, 0, 2)
B. corallinus†† Beschin, Busulini, De Angeli & Tessier, 2007
B. februarius†† Müller, 1974b
Eriocheir† De Haan, 1835 (3, 1, 0)
E. japonica† (De Haan, 1835)
Hemigrapsus† Dana, 1851f (12, 2, 0)
H. nudus† (Dana, 1851f)
H. oregonensis† (Dana, 1851f)
Utica† White, 1847d (7, 2, 0)
U. haswelli† Wintle, 1886
U. yarraensis† Wintle, 1886
Xenograpsidae Ng, Davie, Schubart & Ng, 2007
No fossil representatives
Fossil grapsoid genera not assigned to family

Daragrapsus†† Müller & Collins, 1991 (0, 0, 1)
D. trispinosus Müller & Collins, 1991 (type)
Daranyia†† Lőrenthey, 1901 (0, 0, 2)
D. fabianii Di Salvo, 1933
D. granulata Lőrenthey, 1901a (type)
Pseudodaranyia†† Tessier, Beschin, Busulini & De Angeli, 1999 (0, 0, 1)
P. carinata Tessier et al., 1999 (type)
Grapsus dubius Desmarest, 1822

Incertae sedis

Anaglyptus†† A. Milne-Edwards, 1873a (0, 0, 1)
A. delfortrei A. Milne-Edwards, 1873a (type)
Aulacopodia†† Bosquet, 1854 (0, 0, 1)
A. riemsdyki Bosquet, 1854 (type)
Brome†† Münster, 1839 (0, 0, 1)
B. ventrosa Münster, 1839 (type)
Cardirhynchus†† Schlüter, 1862 (0, 0, 1)
C. spinosus Schlüter, 1862 (type)
Eurycarpus†† Schlüter *in* Schlüter & Von der Marck, 1868 (0, 0, 1)
E. nanodactylus Schlüter *in* Schlüter & Von der Marck, 1868 (type)
Homelys†† Von Meyer, 1862 (0, 0, 1) (Palaemonidae?)
H. minor Von Meyer, 1862 (type)
Lambropsis†† Lőrenthey, 1909b (0, 0, 1)
L. wanneri Lőrenthey, 1909b (type)
Liocaris†† Van Straelen, 1924 [imprint 1925] (0, 0, 1)
L. quadratus Van Straelen, 1924 [imprint 1925] (type)
Lissopsis†† Fritsch & Kafka, 1887 (0, 0, 1)
L. transiens Fritsch & Kafka, 1887 (type)
Lupeites†† Fritsch & Kafka, 1887 (0, 0, 1)
L. granulatus Fritsch & Kafka, 1887 (type)
Microcorystes†† Lőrenthey *in* Lőrenthey & Beurlen, 1929 (0, 0, 1) (junior homonym of *Microcorystes* Fritsch, 1893; Lőrenthey *in* Lőrenthey & Beurlen, 1929, specimen is lost according to Glaessner, 1969)
M. latifrons Lőrenthey *in* Lőrenthey & Beurlen, 1929 (type)
Oedisoma†† Bell, 1858 (0, 0, 1)
O. ambiguum Bell, 1858 (type)
Olinaecaris†† Van Straelen, 1924 [imprint 1925] (0, 0, 1)
O. carinatus Van Straelen, 1924 [imprint 1925] (type)
Phalangites†† Münster, 1839 (0, 0, 1) (possibly larva of *Palinurina*)
P. priscus Münster, 1839 (type)
Psammograpsus†† A. Milne-Edwards, 1860 (0, 0, 1)
P. parisiensis A. Milne-Edwards, 1860 (type)
Pseudomicippe†† Pelseneer, 1886c (0, 0, 1)
P. granulosa Pelseneer, 1886c (type)
Roemerus†† Bishop, 1983a (0, 0, 1)
R. robustus Bishop, 1983a (type)
Secretanella†† Guinot & Tavares, 2001 (0, 0, 1)

S. arcuata (Secretan, 1964) (type)
Tiche†† Von der Marck, 1863 (0, 0, 1)
T. astaciformis Von der Marck, 1863 (type)
Vanua†† Rathbun, 1945 (0, 0, 1)
V. linearius Rathbun, 1945 (type)
Wulaicarcinus†† Hu & Tao, 1996 (0, 0, 1)
W. micronus Hu & Tao, 1996 (type)

**Hasaracancer* was originally referred to the Raninidae (Jux, 1971). Based upon examination of the illustrations, the genus was later moved to the Necrocarcinidae (Schweitzer, Feldmann et al., 2003). Examination of casts of the holotype (BSP 1988III147) of *Hasaracancer cristatus* Jux, 1971, the sole species, deposited in Bayerische Staatssammlung für Paläontologie und Geologie, Munich, suggests that it is more likely a member of the Coleidae.

**Examination of the holotype of *Sagittiformosus carabus* Bishop, 1988b (USNM 418277) indicates that it is a broken carapace referable to *Cenomanocarcinus*, due to its possession of diagnostic rows of tubercles and the Y-shaped pattern of protogastric and mesogastric regions (Schweitzer, Feldmann et al., 2003).

REFERENCES

ADAMS, A. & A. WHITE, 1848. Crustacea, Part I. In: A. ADAMS (ed.), The Zoology of the Voyage of H. M. S. Samarang; under the command of Captain Sir Edward Belcher, C.B., F.R.A.S., F.G.S., during the years 1843-1846: 1-66, pls. 1-13. (Reeve and Benham, London).

— — & — —, 1849. Crustacea, Part II. In: A. ADAMS (ed.), Zoology of the Voyage of H. M. S. Samarang; under the command of Captain Sir Edward Belcher, C.B., F.R.A.S., F.G.S., during the years 1843-1846: 33-67, pls. 7-13. (Reeve, Benham, and Reeve, London).

AGUIRRE[-]URRETA, M. B., 1982. Crustáceos decápodos Barremianos de la región del Tucu-Tucu, Provincia de Santa Cruz. Ameghiniana, **14** (3-4): 303-317.

— —, 1989. The Cretaceous decapod Crustacea of Argentina and the Antarctic Peninsula. Palaeontology, **32**: 499-552.

— —, 1990. Nuevos crustáceos decápodos (Brachyura y Anomura) en el Terciario de Patagonia, Argentina. V Congreso Argentino de Paleontología y Bioestratigrafía, Actas I: 149-154.

— —, 1992. Tertiary freshwater Decapoda (Crustacea: Parastacidae) from the Ñirihuau Basin, Patagonia, Argentina. Journal of Paleontology, **66**: 817-825.

AGUIRRE-URRETA, M. B., L. A. BUATOIS, G. C. B. CHERNOGLASOV & F. A. MEDINA, 1990. First Polychelidae (Crustacea, Palinura) from the Jurassic of Antarctica. Antarctic Science, **2**: 157-162.

AGUIRRE-URRETA, M. B., S. MARENSSI & S. SANTILLANA, 1995. A new Eocene crab (Crustacea, Decapoda) from Seymour Island, Antarctica. Antarctic Science, **7** (3): 277-281.

AHYONG, S. T., 2006. Phylogeny of the clawed lobsters (Crustacea: Decapoda: Homarida). Zootaxa, **1109**: 1-14.

AHYONG, S., 2009. The Polychelidan lobsters: phylogeny and systematics (Polychelida: Polychelidae). In: J. W. MARTIN, K. A. CRANDALL & D. L. FELDER (eds.), Decapod Crustacean Phylogenetics: 369-396. (Crustacean Issues 18, CRC Press, Taylor & Francis Group, New York).

ALBERTI, F. VON *see* VON ALBERTI, F.

ALBRECHT, H. VON *see* VON ALBRECHT, H.

ALCOCK, A., 1896. Material for a carcinological fauna of India, No. 2. The Brachyura Oxystomata. Journal of the Asiatic Society of Bengal, **65** (2): 134-296.

— —, 1898. The family Xanthidae: the Brachyura Cyclometopa, Part I: Materials for a carcinological fauna of India, No. 3. Journal of the Asiatic Society of Bengal, **67**: 67-233.

— —, 1899. Materials for a carcinological fauna of India, No. 4. The Brachyura Cyclometopa. Part II. A revision of the Cyclometopa with an account of the families Portunidae, Cancridae and Corystidae. Journal of the Asiatic Society of Bengal, **68** (2) (1): 1-104.

— —, 1900a. Materials for a carcinological fauna of India, No. 5. The Brachyura Primigenia or Dromiacea. Journal of the Asiatic Society of Bengal, **68** (2) (3): 123-169.

— —, 1900b. Materials for a carcinological fauna of India, No. 6. The Brachyura Catometopa, or Grapsoidea. Journal of the Asiatic Society of Bengal, **69** (2) (3): 279-456.

ALCOCK, A. W., 1910. On the classification of the Potamonidae (Thelphusidae). Records of the Indian Museum, **5** (4): 253-261.

ALIEV, O. B. & R. A. ALIEV, 1980. Burrowing shrimp from the Upper Cretaceous of the Malyi Kavkaz Mountains, USSR. Paleontologicheski Zhurnal, **2**: 141-143.

ALLASINAZ, A., 1987. Brachyura Decapoda oligocenici (Rupeliano) del Bacino Ligure Piemontese. Bollettino del Museo Regionale di Scienze Naturali, Torino, **5** (2): 509-566, pls. 1-7.

ALVAREZ, F., T. M. ILIFFE & J. L. VILLALOBOS, 2006. Macromaxillocarididae, a new family of stenopodidean shrimp from an anchialine cave in the Bahamas, with the description of *Macromaxillocaris bahamaensis*, n. gen., n. sp. Journal of Crustacean Biology, **26**: 366-378.

AMATI, L., R. M. FELDMANN & J.-P. ZONNEVELD, 2004. A new family of Triassic lobsters (Decapoda: Astacidea) from British Columbia and a cladistic analysis of the infraorders Astacidea and Palinura. Journal of Paleontology, **78** (1): 150-168.

AMMON, L. VON *see* VON AMMON, L.

ANDERSON, J. L. & R. M. FELDMANN, 1995. *Lobocarcinus lumacopius* (Decapoda: Cancridae), a new species of cancrid crab from the Eocene of Fayum, Egypt. Journal of Paleontology, **69**: 922-932.

ANGELI, A. DE *see* DE ANGELI, A.

ANNANDALE, N. & S. KEMP, 1913. The Crustacea of Lake Tiberias. Journal and Proceedings of the Asiatic Society of Bengal, (n.s.) **9**: 241-258, pls. 12-14.

ARCHIAC, A. D' *see* D'ARCHIAC, A.

ARTAL, P., 2007. *Uca miocenica* (Crustacea, Decapoda), nueva especie del Mioceno de la Prov. de Barcelona (Cataluña, España). Scripta Musei Geologici Seminarii Barcinonensis, (Palaeontologica) **6**: 3-18.

ARTAL, P., B. VAN BAKEL & J. CASTILLO, 2006. *Retropluma* Gill, 1894 (Crustacea, Decapoda) from the Eocene of the eastern Pyrenees (Spain, France). Cainozoic Research, **5** (1-2): 65-71.

ARTAL, P. & J. CASTILLO, 2005a. *Cyrtorhina ripacurtae* n. sp. (Crustacea, Decapoda, Raninidae), primera cita del género en el Eoceno inferior español. Batalleria, **12**: 33-38.

— — & — —, 2005b. *Periacanthus ramosus* (Crustacea, Decapoda), nueva especie del Eoceno inferior de Huesca (Aragón, España). Batalleria, **12**: 39-44.

ARTAL, P. & A. GILLES, 2007. New Miocene crabs from Pignan (southeast France). Memorie della Società italiana di Scienze naturali e del Museo civico di Storia naturale in Milano, **35** (2): 8-11.

ARTAL, P., D. GUINOT, B. VAN BAKEL & J. CASTILLO, 2008. Ibericancridae, a new dakoticancrid family (Decapoda, Brachyura, Podotremata) from the Upper Campanian (Upper Cretaceous) of Spain. Zootaxa, **1907**: 1-27.

ASHBY, E. & B. C. COTTON, 1936. South Australia fossil chitons. Records of the South Australian Museum, **5**: 509-512.

ASLANOVA, S. M. & J. D. DSCHAFAROVA, 1975. Some finds of Decapoda from the Tertiary deposits of Azerbaijan. Akademia Nauk Azerbaidzhanskoi Soviet Socialist Republic Doklady, **31** (7): 41-46.

ASSMANN, P., 1927. Die Decapodenkrebse des deutschen Muschelkalks. Königliche Preussischen Geologische Landesanstalt Jahrbuch, Berlin, **48**: 332-356.

AVNIMELECH, M., 1961. *Cenomanocarcinus*, a decapod crab from the upper Cenomanian of Jerusalum. The Bulletin of the Research Council of Israel, (G) **10** (1-2): 1-3.

BABA, K., E. MACPHERSON, G. C. B. POORE, S. T. AHYONG, A. BERMUDEZ, P. CABEZAS, C.-W. LIN, M. NIZINSKI, C. RODRIGUES & K. E. SCHNABEL, 2008. Catalogue of squat lobsters of the world (Crustacea: Decapoda: Anomura – families Chirostylidae, Galatheidae and Kiwaidae). Zootaxa, **1905**: 1-220.

BABA, K. & A. B. WILLIAMS, 1998. New Galatheiodea (Crustacea, Decapoda, Anomura) from hydrothermal systems in the west Pacific Ocean: Bismarck Archipelago and Okinawa Trough. Zoosystema, **20**: 143-156.

BACHMAYER, F., 1950. Neue Dekapoden aus dem österreichischen Tertiär. Annalen des Naturhistorischen Museums in Wien, **57**: 133-140, pl. 1.

— —, 1953 [imprint 1951]. Die Decapodenfauna des tortonischen Leithakalkes von Deutsch-Altenburg (Niederösterreich). Mitteilungen der Geologischen Gesellschaft in Wien, **44** [for 1951]: 237-262, pls. 1-5.

— —, 1953a. Zwei neue Crustaceen-Arten aus dem Helvet-Schlier von Ottnang (Oberösterreich). Annalen des Naturhistorischen Museums in Wien, **59**: 137-140, pl. 5.

— —, 1953b. *Goneplax gulderi*, eine neue Crustaceen-Species aus dem tortonischen Tegel des Wiener-Beckens. Paläontologische Zeitschrift, **27** (3/4): 143-145, pl. 9.

— —, 1954a. Zwei bemerkenswerte Crustaceen-Funde aus dem Jungtertiär des Wiener Beckens. Sitzungsberichte der Österreichischer Akademie der Wissenschaften in Wien, (Mathematisch-naturwissenschaftliche Klasse, I) **163** (1, 2): 63-70, pl. 1.

— —, 1954b. *Palaeopalinurus glaessneri* n. g. n. sp. aus dem Oberjura-Kalk von Stramberg in Mähren. Neues Jahrbuch für Geologie und Paläontologie, Abhandlungen, Stuttgart, **99** (2): 147-152.

— —, 1959. Neue Crustaceen aus dem Jura von Stramberg (ČSR). Sitzungsberichte der Österreichischen Akademie der Wissenschaften, (Mathematisch-naturwissenschaftliche Klasse, I) **168** (10): 937-944, pls. 1, 2.

— —, 1962. Die Calappiden (Crustacea, Decapoda) aus den tortonischen Ablagerungen des Wiener Beckens. Annalen des Naturhistorischen Museums in Wien, **65**: 39-46, pls. 1-3.

BACHMAYER, F. & E. MALZAHN, 1983. Der erste Nachweis eines decapoden Krebses im niederrheinischen Kupferschiefer. Annalen des Naturhistorischen Museums in Wien, **85** (A): 99-106.

BACHMAYER, F. & B. MARKOVIĆ, 1955. Ein bemerkenswerter Crustaceen-Fund: *Linuparus petkovici* nov. spec. aus Kreideschichten von Jugoslavien. Zbornik Radova Geoloskog Institut Beograd, Belgrad, **8**: 383-392.

BACHMAYER, F. & M. MOHANTI, 1973. Neue fossile Krebse aus dem Tertiär von Ost-Indien. Annalen des Naturhistorischen Museums in Wien, **77**: 63-67, pls. 1-3.

BACHMAYER, F. & R. MUNDLOS, 1968. Die tertiären Krebse von Helmstedt bei Braunschweig, Deutschland. Annalen des Naturhistorischen Museums in Wien, **72**: 649-692.

BACHMAYER, F. & A. NOSAN, 1959. Ein bemerkenswerte Crustaceenfund aus Gračšiče bei Kubed (Nordistrien). Geologija-Razprave in Porošila, **5**: 80-85.

BACHMAYER, F. & G. PRETZMANN, 1971. Krebsreste aus den altpliozänen Süßwasserablagerungen des Eichkogels bei Mödling, Niederösterreich. Annalen des Naturhistorischen Museums in Wien, **75**: 283-291, pl. 1.

BACHMAYER, F. & R. F. RUTSCH, 1962. Brachyurenfunde (Crustacea) aus der miozänen Meeresmolasse der Schweiz. Eclogae Geologicae Helvetiae, **55** (2): 675-688, pls. 1-3.

BACHMAYER, F. & A. TOLLMANN, 1953. Die Crustaceen-Fauna aus dem tortonischen Leithakalk (Steinbrüche der Firma Fenk) bei Groß-Höflein im Burgenland. In: Kober-Festschrift, Skizzen zum Antlitz der Erde: 308-314, pl. 1. (Universität Wien, Vienna).

BACHMAYER, F. & R. WAGNER, 1957. *Geryon heimertingensis* n. sp., eine Krabbe aus dem Chattian der Erdölbohrung Heimertingen (bayerische Vorlandmolasse). Paläontologische Zeitschrift, **31** (1/2): 99-102.

BAKEL, B. W. M. VAN *see* VAN BAKEL, B. W. M.

BALL, H. W., 1960. Upper Cretaceous Decapoda and Serpulidae from James Ross Island, Graham Land. Falkland Islands Dependencies Survey Scientific Reports, **24**: 1-30.

BALSS, H., 1913. Über fossile Galatheiden. Centralblatt für Mineralogie, **1913**: 155-160.

— —, 1924. Ostasiatische Decapoden. V. Die Oxyrhynchen und Schlußteil (Geographische Übersicht der Decapoden Japans). Archiv für Naturgeschichte, (A) **90** (5): 20-84, figs. 1, 2, pl. 1.

— —, 1932. Über einige systematisch interessante Xanthidae (Crustacea, Decapoda, Brachyura) der Harmsschen Reisen nach dem Sundaarchipel. Zeitschrift fur Wissenschaftliche Zoologie, **142** (2): 510-519.

BARKER-WEBB, P. & S. BERTHELOT, 1836-1844. Histoire naturelle des Iles Canaries, **2** (2), Contenant la zoologie: 1-900 [approx. number of pp.]. (Béthune, Paris).

BARNES, R. S. K., 1967. The Macrophthalminae of Australasia; with a review of the evolution and morphological diversity of the type genus *Macrophthalmus* (Crustacea: Brachyura). Transactions of the Zoological Society of London, **31** (2): 195-262, figs. 1-16, pls. 1-4.

BARNOLAS CORTINAS, A., 1973. *Dromilites vicensis* n. sp., nuevo Braquiuro del Eoceno marino de Cataluña. Instituto de Investigaciones Geológicas, **28**: 5-13.

BATE, C. S., 1859. On the importance of an examination of the structure of the integument of Crustacea in the determination of doubtful species. – Application to the genus *Galathea*, with the description of a new species of that genus. Journal of the Proceedings of the Linnean Society of London, **3**: 1-4.

— —, 1881. On the Penaeidea. Annals and Magazine of Natural History, (5) **8**: 169-196.

— —, 1884. *Archaeoastacus* (*Eryon*) *Willemoesii*, a new genus and species of Eryonidae. Geological Magazine, **1**: 307-310.

— —, 1888. Report on the Crustacea Macrura collected by H. M. S. "Challenger" during the years 1873-1876. Reports on the Scientific Results of the Voyage of H. M. S. Challenger, (Zoology) **24**: 1-942. (Published by Order of Her Majesty, London).

BEAN, W., 1839. A catalogue of the fossils found in the Cornbrash Limestone of Scarborough. Magazine of Natural History, (new series) **3**: 57-62.

BEETS, C., 1950. On fossil brachyuran crabs from the East Indies. Verhandelingen van het Koninklijk Nederlands Geologisch Mijnbouwkundig Genootschap, (Geologische serie) **15**: 349-354, 1 pl.

BEIKIRCH, D. W. & R. M. FELDMANN, 1980. Decapod crustaceans from the Pflugerville Member, Austin Formation (Late Cretaceous: Campanian) of Texas. Journal of Paleontology, **54**: 309-324.

BELL, M. C., F. REDANT & I. TUCK, 2006. *Nephrops* species. In: B. F. PHILLIPS (ed.), Lobsters: biology, management, aquaculture and fisheries: 412-462. (Blackwell Publishing, Oxford, U.K.).

BELL, T., 1836. Some account of the Crustacea of the coasts of South America, with descriptions of new genera and species, founded principally on the collections obtained by Mr. Cuming and Mr. Miller. (Tribus 1, Oxyrhynchi). Proceedings of the Zoological Society of London, **3** [1835]: 169-173.

— —, 1844. On a fossil Crustacean from New Holland. Annals and Magazine of Natural History, **14**: 455.

— —, 1845. On the *Thalassina Emeryi*, a fossil Crustacean forwarded by Mr. W. S. Macleay from New Holland. Quarterly Journal of the Geological Society, **1**: 93-94.

— —, 1850. Notes on the Crustacea of the Chalk Formation. In: F. DIXON (ed.), The geology and fossils of the Tertiary and Cretaceous formations of Sussex: 344-345, pl. 38. (Brown, Green & Longmans, London).

— —, 1855. A monograph of the Leucosiadae, with observations on the relations, structure, habits and distribution of the family: a revision of the generic characters and descriptions of new genera and species, Hora Carcinologicae, or notices of Crustacea, I. Transactions of the Linnean Society, London, **21**: 277-314, pls. 30-34.

— —, 1858. A monograph of the fossil malacostracous Crustacea of Great Britain, Pt. I, Crustacea of the London Clay. Monograph of the Palaeontographical Society, London, **10** [1856]: i-viii, 1-44, 11 pls.

— —, 1863. A monograph of the fossil malacostracous Crustacea of Great Britain, Pt. II, Crustacea of the Gault and Greensand. Palaeontographical Society Monograph, London: 1-40, 11 pls.

BENEDEN, P. J. VAN *see* VAN BENEDEN, P. J.

BENEDICT, J. E., 1904. A new genus and two new species of crustaceans of the family Albuneidae from the Pacific Ocean; with remarks on the probable use of the antennule in *Albunea* and *Lepidopa*. Proceedings of the United States National Museum, **27**: 621-625.

BERGLUND, R. E. & R. M. FELDMANN, 1989. A new crab, *Rogueus orri* n. gen. and sp. (Decapoda: Brachyura), from the Lookingglass Formation (Ulatisian Stage: Lower Middle Eocene) of southwestern Oregon. Journal of Paleontology, **63**: 69-73.

BERGLUND, R. E. & J. L. GOEDERT, 1992. A new species of *Cancer* (Decapoda: Brachyura) from the Miocene Astoria Formation in Washington. Burke Museum Contributions in Anthropology and Natural History, **9**: 1-11.

— — & — —, 1996. A new crab (Brachyura: Cancridae) from Lower Miocene rocks of the northwestern Olympic Peninsula, Washington. Journal of Paleontology, **70** (5): 830-835.

BERVILLE, P. DE *see* DE BERVILLE, P.

BESCHIN, C. & A. DE ANGELI, 1984. Nuove forme fossili di Anomura Hippidae *Albunea cuisiana* sp. n. e *Albunea lutetiana* sp. n. Lavori – Società Veneziana di Scienze Naturali, **9** (1): 93-107.

— — & — —, 2003. *Spinipalicus italicus*, nuovo genere e specie di Palicidae (Crustacea, Decapoda) dell'Eocene del Vicentino (Italia settentrionale). Studi e Ricerche – Associazione Amici del Museo – Museo Civico "G. Zannato", Montecchio Maggiore (Vicenza), **2003**: 7-12.

— — & — —, 2004. Nuovi brachiuri eocenici del Monti Lessini Vicentini (Italia nordorientale). Studi e Ricerche – Associazione Amici del Museo – Museo Civico "G. Zannato", Montecchio Maggiore (Vicenza), **11**: 13-22.

BESCHIN, C., A. DE ANGELI & A. CHECCHI, 2001. Crostacei decapodi associate a coralli della "Formazione di Castelgomberto" (Oligocene) (Vicenza-Italia settentrionale). Studi e Ricerche – Associazione Amici del Museo – Museo Civico "G. Zannato", Montecchio Maggiore (Vicenza), **2001**: 13-30.

— —, — — & — —, 2007. Crostacei dell'Eocene Inferiore di Cava "Rossi" di Monte di Malo (Vicenza, Italia settentrionale). Studi e Ricerche – Associazione Amici del Museo – Museo Civico "G. Zannato", Montecchio Maggiore (Vicenza), **14**: 11-24.

BESCHIN, C., A. DE ANGELI, A. CHECCHI & P. MIETTO, 2006. Crostacei del Priaboniano di Priabona (Vicenza – Italia settentrionale). Lavori – Società Veneziana di Scienze Naturali, **31**: 95-112.

BESCHIN, C., A. DE ANGELI, A. CHECCHI & G. ZARANTELLO, 2005. Crostacei di Grola presso Spagnago (Vicenza, Italia settentrionale). Studi e Ricerche – Associazione Amici del Museo – Museo Civico "G. Zannato", Montecchio Maggiore (Vicenza), **12**: 5-35.

BESCHIN, C., A. DE ANGELI & A. GARASSINO, 2001. *Justitia vicetina* sp. nov. (Crustacea, Decapoda) dell'Eocene di Chiampo (Vicenza, Italia settentrionale). Studi Trentini di Scienze Naturali, (Acta Geologica) **76** [1999]: 89-97.

BESCHIN, C., A. BUSULINI, A. DE ANGELI & G. TESSIER, 1985. Il genere *Micromaia* Bittner (Crustacea, Decapoda) nel Terziario dell'area dei Berici e dei Lessini, con descrizione di tre nuove specie. Lavori – Società Veneziana di Scienze Naturali, **10**: 97-119.

— —, — —, — — & — —, 1988. Raninidae del Terziario berico-lessineo (Italia settentrionale). Lavori – Società Veneziana di Scienze Naturali, **13**: 155-215.

— —, — —, — — & — —, 1994. I crostacei eocenici della Cava «Boschetto» di Nogarole Vicentino (Vicenza – Italia settentrionale). Lavori – Società Veneziana di Scienze Naturali, **19**: 159-215.

— —, — —, — — & — —, 1996a. *Retroplumoidea* (Crustacea, Brachyura) nel Terziario del Vicentino (Italia settentrionale). Lavori – Società Veneziana di Scienze Naturali, **21**: 83-102.

— —, — —, — — & — —, 1996b. *Eopalicus* nuovo genere di brachiuro (Decapoda) del Terziario Veneto (Italia Settentrionale). Lavori – Società Veneziana di Scienze Naturali, **21**: 75-82.

— —, — —, — — & — —, 2002. Aggiornamento ai crostacei di cava "Main" di Arzignano (Vicenza – Italia settentrionale) (Crustacea, Decapoda). Studi e Ricerche – Associazione Amici del Museo – Museo Civico "G. Zannato", Montecchio Maggiore (Vicenza), **2002**: 7-28.

— —, — —, — — & — —, 2004. The Eocene decapod crustacean fauna of the "Main" quarry in Arzignano (Vicenza – NE Italy) with the description of a new species of Raninidae. Lavori – Società Veneziana di Scienze Naturali, **29**: 109-117.

— —, — —, — — & — —, 2007. I Decapodi dell'Eocene Inferiore di Contrada Gecchelina (Vicenza – Italia settentrionale) (Anomura e Brachiura). Museo di Archeologia e Scienze Naturali "G. Zannato", Montecchio Maggiore (Vicenza), **2007**: 9-76.

BESCHIN, C., A. BUSULINI, A. DE ANGELI, G. TESSIER & S. UNGARO, 1991. Due nuovi generi di Raninidae dell'Eocene del Veneto (Italia). Lavori – Società Veneziana di Scienze Naturali, **16**: 187-212.

BESCHIN, C. & A. GARASSINO, 1999. *Penaeus vanzii* n. sp. e *Penaeus sorbinii* n. sp. (Crustacea, Decapoda) dell'Oligocene della Valle del Ponte (Laverda) e di Salcedo (Vicenza, N. Italia). Atti della Società italiana di Scienze naturali e del Museo civico di Storia naturale in Milano, **140** (2): 189-208.

BEURLEN, K., 1925. Über Brachyuren- und Anomurenreste des schwäbischen Jura. Neues Jahrbuch für Mineralogie, (Beilageband) **52** (3): 464-532, 2 figs.

— —, 1928a. Die Decapoden des schwäbischen Jura mit Ausnahme der aus den oberjurassischen Plattenkalken stammenden. Palaeontographica, **70**: 115-278, pls. 6-8.

— —, 1928b. Die fossilen Dromiaceen und ihre Stammesgeschichte. Paläontologische Zeitschrift, **10**: 144-183.

— —, 1929. Untersuchungen über Prosoponiden. Centralblatt für Mineralogie, Geologie, und Paläontologie, (B, Geologie und Paläontologie) **1929**: 125-142.

— —, 1930a. Vergleichende Stammesgeschichte, Grundlagen, Methoden, Probleme unter besonderer Berücksichtigung der höheren Krebse. Fortschritte in der Geologie und Paläontologie, **8**: 317-586.

— —, 1930b. Nachträge zur Decapodenfauna des schwäbischen Jura. I. Neue Decapodenfunde aus dem Posidonienschiefer von Holzmaden. Neues Jahrbuch für Mineralogie, Geologie und Paläontologie, (B) **64**: 219-234.

— —, 1932. Brachyurenreste aus dem Lias von Bornholm mit Beiträgen zur Phylogenie und Systematik der brachyuren Dekapoden. Paläontologische Zeitschrift, **14**: 52-66.

— —, 1933. Crustacea Decapoda aus den Tendaguru-Schichten. Palaeontographica, **7** (2-2): 89-94.

— —, 1938. Algunos fosiles cretacios (vermes, equinods, crustaceos) de la Cordillera Oriental. Estudios Geologico y Paleontologico sobre la Cordillera de Colombia, **3**: 128-136.

— —, 1939. Neue Dekapoden-Krebse aus dem ungarischen Tertiär. Palaontologische Zeitschrift, **21**: 135-161, pl. 7.
— —, 1944. Neue Reste von *Proeryon* (Crustacea Decapoda, Eryonidea). Neues Jahrbuch für Geologie und Paläontologie, **88**: 374-384.
— —, 1950. Alguns restos de Crustáceos Decápodes d'água doce fósseis no Brasil. Anais da Academia Brasiliera de Ciencias, **22** (4): 453-459.
— —, 1958a. Contribuição à paleontologia do estado do Para: Crustáceos Decápodos da Formação Pirabas. Boletim do Museu Paraense Emilio Goeldi, (nova série) (Geologia) **5**: 1-48, pls. 1-4.
— —, 1958b. Dois crustaceos do Cretáceo superior do nordeste do Brasil (Decapoda, Brachyura). Boletim do Museu Nacional, Rio de Janeiro, (nova série) **26**: 1-23.
— —, 1962. O género *Callianassa* nas formações Cretácicas de Pernambuco. Arquivos de Geologia, Universidad do Recife, **2**: 1-10.
— —, 1965. Crustáceos Decápodos na Formação Riachuelo (Cretáceo-Sergipe). Anais da Academia Brasiliera de Ciencias, **37** (2): 267-272.
BIFFAR, T. A., 1970. Three new species of callianassid shrimp (Decapoda, Thalassinidea) from the western Atlantic. Proceedings of the Biological Society of Washington, **83** (3): 35-49.
BILL, PH. C., 1914. Ueber Crustaceen aus dem Voltziensandstein des Elsasses. Mitteilungen Geologische Landesanstalt Elsass-Lotharingen, Strassburg, **8**: 289-338.
BINKHORST, J. T. VAN *see* VAN BINKHORST, J. T.
BIRSHTEIN, Y. A., 1956. The decapod Crustacea of the Ferganian Paleogene. Bulletin of the Moscow Society for Natural Research, (Geological Section) **31** (1): 74-96. (Israel Program for Scientific Translations, Jerusalem, 1961).
— —, 1958. Ein Vertreter der ältesten Ordo der Crustacea Decapoda *Protoclitiopsis antiqua* gen. nov. sp. nov. aus dem Permo West-Sibiriens. Doklady Akademii Nauk, SSSR, **122**: 477-480.
BISHOP, G. A., 1973. *Homolopsis dawsonensis*: a new crab (Crustacea, Decapoda) from the Pierre Shale (Upper Cretaceous, Maastrichtian) of Cedar Creek Anticline, eastern Montana. Journal of Paleontology, **47** (1): 19-20, pl. 1.
— —, 1976. *Ekalakia lamberti* n. gen., n. sp. (Crustacea, Decapoda) from the Upper Cretaceous Pierre Shale of eastern Montana. Journal of Paleontology, **50**: 398-401.
— —, 1978. Two new crabs, *Sodakus tatankayotankaensis* n. gen., n. sp. and *Raninella oaheensis* n. sp. (Crustacea, Decapoda), from the Upper Cretaceous Pierre Shale of South Dakota. Journal of Paleontology, **52**: 608-617.
— —, 1982. *Homolopsis mendryki*: a new fossil crab (Crustacea, Decapoda) from the Late Cretaceous *Dakoticancer* assemblage, Pierre Shale (Maastrichtian) of South Dakota. Journal of Paleontology, **56** (1): 221-225.
— —, 1983a. Fossil decapod crustaceans from the Lower Cretaceous, Glen Rose Limestone of central Texas. Transactions of the San Diego Society of Natural History, **20**: 27-55.
— —, 1983b. Two new species of crabs, *Notopocorystes* (*Eucorystes*) *eichhorni* and *Zygastrocarcinus griesi* (Decapoda: Brachyura) from the Bearpaw Shale (Campanian) of north-central Montana. Journal of Paleontology, **57** (5): 900-910.
— —, 1983c. Fossil decapod Crustacea from the Late Cretaceous Coon Creek Formation, Union County, Missisippi. Journal of Crustacean Biology, **3**: 417-430.
— —, 1985a. Fossil decapod crustaceans from the Gammon Ferruginous Member, Pierre Shale (Early Campanian), Black Hills, South Dakota. Journal of Paleontology, **59** (3): 605-624.
— —, 1985b. A new crab, *Prehepatus harrisi* (Crustacea Decapoda), from the Coon Creek and Prairie Bluff formations, Union County, Mississippi. Journal of Paleontology, **59**: 1028-1032.

——, 1985c. A new crab, *Eumunidopsis cobbani* n. sp. (Crustacea, Decapoda) from the Pierre Shale (Early Maastrichtian) of Colorado. Journal of Paleontology, **59** (3): 601-604.

——, 1986a. Two new crabs, *Parapaguristes tuberculatus* and *Palaeoxantho libertiensis*, from the Prairie Bluff Formation (Middle Maastrichtian), Union County, Mississippi, U.S.A. Proceedings of the Biological Society of Washington, **99**: 604-611.

——, 1986b. A new crab, *Zygastrocarcinus cardsmithi* (Crustacea, Decapoda), from the lower Pierre Shale, southeastern Montana. Journal of Paleontology, **60**: 1097-1102.

——, 1987. *Dromiopsis kimberlyae*, a new Late Cretaceous crab from the Pierre Shale of South Dakota. Proceedings of the Biological Society of Washington, **100**: 35-39.

——, 1988a. A new crab, *Seorsus wadei*, from the Late Cretaceous Coon Creek Formation, Union County, Mississippi. Proceedings of the Biological Society of Washington, **101**: 72-78.

——, 1988b. New fossil crabs, *Plagiophthalmus izetti*, *Latheticocarcinus shapiroi*, and *Sagittiformosus carabus* (Crustacea, Decapoda) from the Western Interior Cretaceous, U.S.A. Proceedings of the Biological Society of Washington, **101**: 375-381.

——, 1988c. Two crabs, *Xandaros sternbergi* (Rathbun, 1926) n. gen., and *Icriocarcinus xestos* n. gen., n. sp., from the Late Cretaceous of San Diego County, California, U.S.A., and Baja California Norte, Mexico. Transactions of the San Diego Society of Natural History, **21**: 245-257.

——, 1991. *Xanthosia occidentalis* Bishop, 1985, and *Xanthosia spinosa*, new species, two late Cretaceous crabs from the Pierre Shale of the Western Interior. Journal of Crustacean Biology, **11** (2): 305-314.

——, 1992a. A new Cretaceous crab *Diaulax millerae* (Crustacea: Decapoda), from the northern Atlantic Coastal Plain, U.S.A. Proceedings of the Biological Society of Washington, **105**: 555-561.

——, 1992b. Two new crabs, *Homolopsis williamsi* and *Homolopsis centurialis* (Crustacea; Decapoda), from the Western Interior Cretaceous of the United States. Proceedings of the Biological Society of Washington, **105**: 55-66.

BISHOP, G. A. & N. A. BRANNEN, 1992. *Homolopsis pikae*, new species (Decapoda), a crab from the Cretaceous of Texas. Journal of Crustacean Biology, **12** (2): 317-323.

BISHOP, G. A. & B. T. PALMER, SR., 2006. A new genus and species of crab from the bryozoan bioherms of the Eocene Santee Limestone; South Carolina, U.S.A. Revista Mexicana de Ciencias Geológicas, **23** (3): 334-337.

BISHOP, G. A. & R. PORTELL, 1989. A Pliocene crab-sea star association from southwest Florida. Journal of Crustacean Biology, **9** (3): 453-458.

BISHOP, G. A. & A. B. WILLIAMS, 1991. *Necrocarcinus olsonorum*, new species, a crab (Decapoda, Calappidae) from the Cretaceous Carlile Shale (Turonian), Western Interior United States. Journal of Crustacean Biology, **11** (3): 451-459.

—— & ——, 2000. Fossil crabs from Tepee Buttes, submarine seeps of the late Cretaceous Pierre Shale, South Dakota and Colorado, U.S.A. Journal of Crustacean Biology, **20** (special number) (2): 285-300.

BITTNER, A., 1875. Die Brachyuren des vicentinischen Tertiärgebirges. Denkschriften der Kaiserlichen Akademie der Wissenschaften, (Mathematisch-naturwissenschaftliche Klasse) **34**: 63-105, pls. 1-5.

——, 1882. In: H. ABICH (ed.), Geologische Forschungen in den Kaukasusländern. II. Geologie des Armenischen Hochlandes. I. Westhälfte: 1-488, 19 pls. (A. Hölder, Vienna).

——, 1883. Neue Beiträge zur Kenntnis der Brachyuren-Fauna des Alttertiärs von Vicenza und Verona. Denkschriften der Kaiserlichen Akademie der Wissenschaften, (Mathematisch-naturwissenschaftliche Klasse) **46**: 299-316, pl. 1.

— —, 1884. Beiträge zur Kenntnis tertiärer Brachyuren-Faunen. Denkschriften der Kaiserlichen Akademie der Wissenschaften, (Mathematisch-naturwissenschaftliche Klasse) **48** (1): 15-30, pls. 1, 2.

— —, 1886. Neue Brachyuren des Eozäns von Verona. Sitzungsberichte der Kaiserlichen Akademie der Wissenschaften in Wien, (Mathematisch-naturwissenschaftliche Klasse) **94** (1): 44-55.

— —, 1893. Decapoden des pannonischen Tertiäre. Sitzungsberichte der Kaiserlichen Akademie der Wissenschaften in Wien, **102**: 10-37, pls. 1, 2.

BLAKE, J. F., 1876. Palaeontology. In: R. TATE & J. F. BLAKE (eds.), The Yorkshire Lias: 243-475. (Van Voorst, London).

BLASCHKE, F., 1911. Zur Tithonfauna von Stramberg in Mähren. Annalen des Kaiserlich-Königlichen Naturhistorischen Museums in Wien, **25** (1, 2): 143-222.

BLOW, W. C., 2003. New brachyuran crabs (Crustacea: Decapoda) from the Upper Pliocene Yorktown Formation of southeastern Virginia. Proceedings of the Biological Society of Washington, **116**: 168-189.

BLOW, W. C. & R. H. BAILEY, 1992. *Chasmocarcinus robertsi*, a new crab species from the Miocene of Virginia, with notes on the genus *Falconoplax* (Crustacea, Decapoda, Goneplacidae). Tulane Studies in Geology and Paleontology, **25**: 175-185.

BLOW, W. C. & R. C. MANNING, 1996. Preliminary descriptions of 25 new decapod crustaceans from the Middle Eocene of the Carolinas, U.S.A. Tulane Studies in Geology and Paleontology, **29** (1): 1-26, pls. 1-5.

— — & — —, 1997. A new genus, *Martinetta*, and two new species of xanthoid crabs from the Middle Eocene Santee Limestone of South Carolina. Tulane Studies in Geology and Paleontology, **30**: 171-180.

— — & — —, 1998. *Eohalimede sandersi*, the correct name for the species described as *Eohalimede saundersi* Blow and Manning, 1997 (Crustacea: Decapoda: Xanthidae). Proceedings of the Biological Society of Washington, **111**: 409.

BÖHM, J., 1891. Die Kreidebildungen des Fürbergs und Sulzbergs bei Siegsdorf in Oberbayern. Palaeontographica, **38**: 1-106, pls. 1-5.

— —, 1911. *Callianassa burckhardti* n. sp. nebst einer Zusammentstellung der fossilen Arten der Gattung *Callianassa*. Zeitschrift der Deutschen Geologischen Gesellschaft, **63**: 37-41.

— —, 1922. Arthropoda. Crustacea. In: K. MARTIN (ed.), Die Fossilien von Java. Sammlungen des Geologischen Reichsmuseums in Leiden, (neue Folge) **1** (2): 521-535, pl. 63.

— —, 1926. Über tertiäre Versteinerungen von den Bogenfelser Diamantfeldern. In: E. KAYSER (ed.), Die Diamantenwüste Südafrikas, **2**: 55-87, pls. 31-34.

— —, 1927. *Raninellopsis Gottschei* nov. gen. n. sp. Jahrbuch der Preussischen Geologischen Landesanstalt und Bergakademie in Berlin, **48**: 563-566, pls. 1, 2.

BOOTH, J. D., 2006. *Jasus* species. In: B. F. PHILLIPS (ed.), Lobsters: biology, management, aquaculture and fisheries: 340-359. (Blackwell Publishing, Oxford, U.K.).

BORRADAILE, L. A., 1902. Marine Crustaceans. I. On varieties. II. Portunidae. In: J. S. GARDINER (ed.), The fauna and geography of the Maldive and Laccadive archipelagoes, **1** (2): 191-208.

— —, 1903. On the classification of the Thalassinidea. Annals and Magazine of Natural History, (7) **12**: 534-551.

— —, 1915. Notes on Carides. Annals and Magazine of Natural History, **15**: 205-213 [205].

BORRISIAK, A., 1904. Sur les restes de Crustacés dans les dépots du Crétacé Inférieur de la Crimée. Bulletin du Comité Géologique de St. Petersbourg, **23**: 403-423, pl. 13.

BOSC, L. A. G., 1802. Histoire naturelle des Crustacés, contenant leur description et leurs moeurs, avec figures dessinées d'après nature, **1**: 1-258, pls. 1-8; **2**: 1-296, pls. 9-18. (De Guilleminet, Paris).

BOSQUET, J., 1854. Les Crustacés fossiles du terrain Crétacé du Limbourg. Verhandelingen van den Commissie van de Geologische Kaart van Nederland, **2**: 12-128.

BOTT, R., 1950. Die Flusskrebse Europas (Decapoda, Astacidae). Abhandlungen der Senckenbergischen Naturforschenden Gesellschaft, **483**: 1-36.

— —, 1965. Die Süsswasserkrabben von Madagaskar (Crustacea, Decapoda). Bulletin du Museum National d'Histoire Naturelle, Paris, (2) **37** (2): 335-350, pls. 1-4.

— —, 1970. Die Süsswasserkrabben von Europa, Asien, Australien und ihre Stammesgeschichte. Eine Revision der Potamoidea und Parathelphusoidea (Crustacea, Decapoda). Abhandlungen der Senckenbergischen Naturforschenden Gesellschaft, Frankfurt, **526**: 1-338, pls. 1-58.

BOUILLÉ, R. DE *see* DE BOUILLÉ, R.

BOURNE, G. C., 1922. On the Raninidae: a study in carcinology. Journal of the Linnean Society of London, (Zoology) **35**: 25-78, pls. 4-7.

BOUVIER, E.-L., 1890. Révision des Cénobites du Muséum. Bulletin de la Société Philomatique de Paris, (8) **2** (4): 143-150.

— —, 1895. Recherches sur les affinités des *Lithodes* et des *Lomis* avec les Paguridés. Annales des Sciences Naturelles, (Zoologie) (7) **18**: 157-213.

— —, 1897. Sur la classification, les origines et la distribution des Crabes de la famille des Dorippidés. Bulletin de la Societé Philomatique de Paris, (8) **9**: 54-70.

— —, 1898. Observations on the crabs of the family Dorippidae. Annals and Magazine of Natural History, (7) **1**: 103-105.

— —, 1899. *Calappa Zurcheri*, Crabe nouveau des terrains miocènes de Panama. Bulletin du Muséum d'Histoire Naturelle, Paris, **5**: 189-192.

BOWDICH, T. E., 1825. Excursions in Madeira and Porto Santo during the autumn of 1823, while on his third voyage to Africa: i-xii, 1-278. (London).

BOYKO, C. B., 2002. A worldwide revision of the recent and fossil sand crabs of the Albuneidae Stimpson and Blepharipodidae, new family (Crustacea, Decapoda, Anomura, Hippoidea). Bulletin of the American Museum of Natural History, **272**: 1-396.

— —, 2004. The problematic *Paralbunea* Hu and Tao, 1996: Homonymy, generic nom. nov., and correct taxonomic placement. Palaeontology, **47**: 929–932.

BRAVI, S., M. G. COPPA, A. GARASSINO & R. PATRICELLI, 1999. *Palaemon vesolensis* n. sp. (Crustacea, Decapoda) from the Plattenkalk of Vesole Mount (Salerno, S Italy). Atti della Società italiana di Scienze naturali e del Museo civico di Storia naturale in Milano, **140** (2): 141-169.

BRAVI, S. & A. GARASSINO, 1998a. New biostratigraphic and palaeoecologic observations on the "Plattenkalk" of the Lower Cretaceous (Albian) of Pietraroia (Benevento, S Italy), and its decapod crustaceans assemblage. Atti della Società italiana di Scienze naturali e del Museo civico di Storia naturale in Milano, **138**: 119-171.

— — & — —, 1998b. "Plattenkalk" of the Lower Cretaceous (Albian) of Petina in the Alburni Mounts (Campania, S Italy), and its decapod crustaceans assemblage. Atti della Società italiana di Scienze naturali e del Museo civico di Storia naturale in Milano, **138**: 89-118.

BRITO, I. M., 1972. Contribuição ao conhecimento dos Crostáceos Decápodos da formação Pirabas. II – O gênero Uca (Brachyura-Ocypodidae). Anais da Academia Brasiliera de Ciencias, **44** (1): 95-98.

BROCCHI, P., 1875. Note sur une nouvelle espèce de Crustacé fossile (*Penaeus libanensis*). Bulletin de la Societé Geologique de France, Paris, **3**: 609-610.

— —, 1877. Description de quelques Crustacés fossiles appartenant à la tribu des Raniniens. Annales de la Societé Géologique, **8**: 1-6, pl. 29.

— —, 1883. Notes sur les Crustacés fossiles des terres tertiaires de la Hongrie. Annales des Sciences Géologiques, (2) **14**: 1-8, pls. 4, 5.

BRODERIP, W., 1835. Crustacea. Proceedings of the Geological Society of London, **2**.

BROOKS, H. K., 1962. The Paleozoic Eumalacostraca of North America. Bulletins of American Paleontology, **44**: 221-338.

BRONN, H. G., 1858. Beiträge zur triasischen Fauna und Flora der bituminösen Schiefer von Raibl. Neues Jahrbuch für Mineralogie, Geologie, und Paläontologie, Stuttgart, **1858**: 1-32.

BRONN, H. G. & F. ROEMER, 1852. Lethaea geognostica, **2** (5). (E. Schweizerbart, Stuttgart).

BRUCE, A. J., 1993. *Kakaducaris glabra* gen. nov., sp. nov., a new freshwater shrimp from the Kakadu National Park, Northern Territory, Australia, Crustacea: Decapoda: Palaemonidae with the designation of a new subfamily Kakaducaridinae. Hydrobiologia, **268**: 27-44.

BURKENROAD, M. D., 1963. The evolution of the Eucarida (Crustacea, Eumalacostraca) in relation to the fossil record. Tulane Studies in Geology, **2**: 3-16.

BUSULINI, A. & C. BESCHIN, 2009. Prima segnalazione di crostacei decapodi nella "Marna di Possagno" (Eocene Superiore – Italia Nordorientale). Lavori – Società Veneziana di Scienze Naturali, **34**: 111-118.

BUSULINI, A., G. TESSIER & C. BESCHIN, 2006. The genus *Phlyctenodes* A. Milne Edwards, 1862 (Crustacea: Decapoda: Xanthidae) in the Eocene of Europe. Revista Mexicana de Ciencias Geológicas, **23** (3): 350-360.

BUSULINI, A., G. TESSIER, C. BESCHIN & A. DE ANGELI, 2003. *Boschettia giampietroi*, nuovo genere e specie di Portunidae (Crustacea, Decapoda) dell'Eocene medio della Valle del Chiampo (Vicenza, Italia settentrionale). Studi e Ricerche – Associazione Amici del Museo – Museo Civico "G. Zannato", Montecchio Maggiore (Vicenza), **2003**: 13-18.

BUSULINI, A., G. TESSIER & M. VISENTIN, 1984. *Titanocarcinus aculeatus* nuova specie di Brachiuro dell'Eocene del Veneto (Crustacea, Decapoda). Lavori – Società Veneziana di Scienze Naturali, **9**: 107-117.

CALMAN, W. T., 1896. On deep-sea Crustacea from the south west of Ireland. Transactions of the Royal Irish Academy, **31**: 1-20.

CAMPBELL, B. M., 1971. New records and new species of crabs (Crustacea: Brachyura) trawled off southern Queensland: Dromiacea, Homolidae, Gymnopleura, Corystoidea, and Oxystomata. Memoirs of the Queensland Museum, **16**: 27-48, pls. 2, 3.

CARRIOL, R.-P., C. DE MUIZON & S. SECRETAN, 1987. Les Crustacés (Cirripedia et Decapoda) du Néogène de la côte Péruvienne. Annales de Paléontologie, (Vert.-Invert.) **73**: 137-164.

CARRIOL, R.-P. & B. RIOU, 1991. Les Dendrobranchiata (Crustacea, Decapoda) du Callovien du La Voulte-sur-Rhône. Annales de Paléontologie, **77** (3): 143-160, pls. 1-4.

CARTER, J., 1872. On *Orithopsis bonneyi*, a new fossil crustacean. Geological Magazine, **9**: 529-532.

— —, 1886. On the decapod crustaceans of the Oxford Clay. Quarterly Journal of the Geological Society, London, **42**: 542-559.

— —, 1898. A contribution to the palaeontology of the decapod Crustacea of England. The Quarterly Journal of the Geological Society of London, **54**: 15-44, pls. I, II.

CARVACHO, A., 1989. *Cancer johngarthi*, n. sp. and *Cancer porteri* (Bell) (Crustacea: Decapoda): comparisons and hypothesis. Proceedings of the Biological Society of Washington, **102**: 613-619.

CASADÍO, S., A. DE ANGELI, R. M. FELDMANN, A. GARASSINO, J. L. HETLER, A. PARRAS & C. E. SCHWEITZER, 2004. New decapod crustaceans (Thalassinidea, Brachyura) from the late Oligocene of Patagonia, Argentina. Annals of Carnegie Museum, **73**: 85-107.

CASADÍO, S., R. M. FELDMANN, A. PARRAS & C. E. SCHWEITZER, 2005. Miocene fossil Decapoda (Crustacea: Brachyura) from Patagonia, Argentina, and their paleoecological setting. Annals of Carnegie Museum, **74**: 151-188.

CASTRO, P., P. K. L. NG & S. T. AHYONG, 2004. Phylogeny and systematics of the Trapeziidae Miers, 1886 (Crustacea: Brachyura), with the description of a new family. Zootaxa, **643**: 1-70.

CAVE, L. DELLE *see* DELLE CAVE, L.

CHACE, F. A., JR., 1937. A correction in crustacean nomenclature. Proceedings of the New England Zoological Club, **16**: 15-16.

— —, 1939. Reports on the scientific results of the first Atlantic Expedition to the West Indies, under the joint auspices of the University of Havana and Harvard University. Preliminary description of one new genus and seventeen new species of decapod and stomatopod Crustacea. Memorias de la Sociedad Cubana de Historia Natural, **13** (1): 31-54.

— —, 1940. The bathypelagic Caridean Crustacea. Plankton of the Bermuda Oceanographic Expeditions. 9. Zoologica, New York, **25**: 117-209.

CHACE, F. A., JR. & R. B. MANNING, 1972. Two new caridean shrimps, one representing a new family, from marine pools on Ascension Island (Crustacea: Decapoda: Natantia). Smithsonian Contributions to Zoology, **131**: 1-18.

CHECCHIA-RISPOLI, G., 1905. I Crostacei dell'Eocene dei dintorni di Monreale in provincia di Palermo. Giornale della Societa Scienze Naturale e Economiche di Palermo, **25**: 309-325, 1 pl.

— —, 1907. Sopra un crostaceo dell'Eocene Medio dei dintorni di Bagheria in Provincia di Palermo. Bollettino della Società Geologica Italiana, **26**: 25-28.

— —, 1917. *Distefania*, nuovo genere di Brachiuri del Cenomaniano della Sicilia. Bollettino della Società Zoologica Italiana, (3) **3** (12): 173-186, pls. 1, 2.

CHEN, H., 1987. On two new species of Leucosiidae (Crustacea: Brachyura) from the Chinese waters. Studia Marina Sinica, **28**: 195-203. [In Chinese with English summary.]

CHERNYSHEV, B., 1930. New Eryonidae from Vilui River. Bulletin of Geology and Prospecting Service, **49**: 375-384.

CHONG, G. & R. FÖRSTER, 1976. *Chilenophoberus atacamensis*, a new decapod Crustacean from the Middle Oxfordian of the Cordillera de Domeyko, northern Chile. Neues Jahrbuch für Geologie und Paläontologie, Monatshefte, **1976** (3): 145-156.

CHRISTOFFERSEN, M. L., 1986. Phylogenetic relationships between Oplophoridae, Atyidae, Pasiphaeidae, Alvinocarididae fam. n., Bresiliidae, Psalidopodidae and Disciadidae (Crustacea Caridea Atyoidea). Bolletim de Zoologia, **10**: 273-281.

CLAUS, C. F. W., 1872. Grundzüge der Zoologie. (Elwert, Marburg und Leipzig).

COBB, J. S. & K. M. CASTRO, 2006. *Homarus* species. In: B. F. PHILLIPS (ed.), Lobsters: biology, management, aquaculture and fisheries: 310-340. (Blackwell Publishing, Oxford, U.K.).

COLLINS, J. S. H., 1969. Some decapod crustaceans from the Lower Cretaceous of Poland and England. Acta Palaeontologica Polonica, **14** (4): 565-575, pls. 1, 2.

— —, 1997. Fossil Homolidae (Crustacea; Decapoda). Bulletin of the Mizunami Fossil Museum, **24**: 51-71.

— —, 2002. A taxonomic review of British decapod Crustacea. Bulletin of the Mizunami Fossil Museum, **29**: 81-92.

COLLINS, J. S. H. & A. J. BARBER, 1998. A new Middle Eocene crab, *Lobocarcinus pentanodosus* sp. nov. (Crustacea, Decapoda) from Doi Doi, Barru, south Sulawesi, Indonesia. Bulletin of the Mizunami Fossil Museum, **25**: 97-101.

COLLINS, J. S. H. & G. BRETON, 2009. New crabs (Crustacea, Decapoda) from the Cenomanian stratotype (Western Paris Basin, France). Bulletin of the Mizunami Fossil Museum, **35**: 43-50.

COLLINS, J. S. H. & I. DIENI, 1995. New decapod crustaceans from the Cenomanian Rudist Limestones of NE Italy. Bulletin of the Mizunami Fossil Museum, **22**: 67-72.

COLLINS, J. S. H. & S. DONOVAN, 1995. A new species of *Necronectes* (Decapoda) from the Upper Oligocene of Antigua. Caribbean Journal of Science, **31** (1-2): 122-127.

— — & — —, 1997. Some new crab records (Crustacea: Decapoda) from the Late Pleistocene Port Morant Formation of southeast Jamaica. Bulletin of the Mizunami Fossil Museum, **24**: 73-77, pl. 14.

— — & — —, 2002. A new species of calappid crab from the Upper Oligocene of Antigua. Caribbean Journal of Science, **38** (1-2): 145-146.

— — & — —, 2004. Decapod crustaceans from the Seroe Doni Formation (Mio-Pliocene) of Aruba, Netherland Antilles. Caribbean Journal of Science, **40** (3): 383-387.

— — & — —, 2005. Eocene crabs (Crustacea Brachyura) from Bonaire, Netherlands Antilles. Scripta Geologica, **129**: 1-12.

— — & — —, 2006. New decapod crustaceans from the Palaeogene of Jamaica. Bulletin of the Mizunami Fossil Museum, **33**: 59-65.

COLLINS, J. S. H., S. K. DONOVAN & H. L. DIXON, 1996. Crabs and barnacles (Crustacea: Decapoda and Cirripedia) from the Late Pleistocene Port Morant Formation of southeast Jamaica. Bulletin of the Mizunami Fossil Museum, **23**: 51-63, pls. 12-19.

COLLINS, J. S. H., S. K. DONOVAN, W. LINDSAY & G. A. STIMPSON, 2001. A new species of portunid crab from the Early Pleistocene Old Pera beds of Jamaica. Proceedings of the Geologists' Association, **112**: 7-12.

COLLINS, J. S. H. & R. H. B. FRAAYE, 1991. *Cancer parvidens*, a new crab (Crustacea, Decapoda) from the Miocene of the Netherlands. Contributions to Tertiary and Quaternary Geology, **28** (1): 1-7, pl. 1.

COLLINS, J. S. H., R. H. B. FRAAYE & J. W. M. JAGT, 1995. Late Cretaceous anomurans and brachyurans from the Maastrichtian type area. Acta Palaeontologica Polonica, **40** (2): 165-210, 12 figs.

COLLINS, J. S. H., R. H. B. FRAAYE, J. W. M. JAGT & P. H. M. VAN KNIPPENBERG, 1997. *Laeviranina hollandica*, a new Late Oligocene (?) raninid crab (Crustacea, Decapoda) from the Netherlands. Tertiary Research, **18**: 1-5.

COLLINS, J. S. H., R. HIGGS & B. CORTITULA, 1994. A new crab, *Costacopluma bifida* (Crustacea, Decapoda) from the Palaeocene of Venezuela. Bulletin of the Mizunami Fossil Museum, **21**: 29-34.

COLLINS, J. S. H. & S. L. JAKOBSEN, 1995. A new crab, *Rogueus robustus*, from the Middle Paleocene of Denmark. Bulletin of the Mizunami Fossil Museum, **22**: 61-65.

— — & — —, 2003. New crabs (Crustacea, Decapoda) from the Eocene (Ypresian/Lutetian) Lillebælt Clay Formation of Jutland, Denmark. Bulletin of the Mizunami Fossil Museum, **30**: 63-96.

COLLINS, J. S. H., Y. KANIE & H. KARASAWA, 1993. Late Cretaceous crabs from Japan. Transactions and Proceedings of the Palaeontological Society of Japan, (new series) **172**: 292-310.

COLLINS, J. S. H. & H. KARASAWA, 1993. The Cretaceous crab, *Pithonoton inflatum* sp. nov. from Hokkaido, Japan. Scientific Reports of the Toyohashi Museum of Natural History, **3**: 17-19.

COLLINS, J. S. H., C. LEE & J. NOAD, 2003. Miocene and Pleistocene crabs (Crustacea, Decapoda) from Sabah and Sarawak. Journal of Systematic Palaeontology, **1** (3): 187-226.

COLLINS, J. S. H., S. F. MITCHELL & S. K. DONOVAN, 2009. A new species of land crab, *Sesarma* Say, 1817 (Decapoda, Brachyura), from the Pleistocene of Jamaica. Scripta Geologica, **138**: 11-21.

COLLINS, J. S. H., R. T. J. MOODY & R. I. SANDMAN, 1999. A new crab and cirripede (Crustacea: Decapoda: Cirripedia) from the Upper Ypresian of central Tunisia. Bulletin of the Mizunami Fossil Museum, **26**: 121-127.

COLLINS, J. S. H. & S. F. MORRIS, 1973. A new crab from the Middle Eocene of Libya. Palaeontology, **16** (2): 283-292, pls. 29, 30.

— — & — —, 1975. A new crab *Costacopluma concava* from the Upper Cretaceous of Nigeria. Palaeontology, **18**: 823-829.

— — & — —, 1976. Tertiary and Pleistocene crabs from Barbados and Trinidad. Palaeontology, **19**: 107-131.

— — & — —, 1978. New Lower Tertiary crabs from Pakistan. Palaeontology, **21**: 957-981.

COLLINS, J. S. H., R. W. PORTELL & S. K. DONOVAN, 2009. Decapod crustaceans from the Neogene of the Caribbean: diversity, distribution and prospectus. Scripta Geologica, **138**: 55-111.

COLLINS, J. S. H. & H. W. RASMUSSEN, 1992. Upper Cretaceous-Lower Tertiary decapod crustaceans from west Greenland. Grønlands Geologiske Undersøgelse, Bulletin, **162**: 1-46.

COLLINS, J. S. H., A. J. ROSS, G. GENZANO & H. MANZIAN, 2006. *Earleria* gen. nov. and *Gabriella* gen. nov., replacement names for *Foersteria* Arai and Brinkmann-Voss, 1980 (Cnidaria, Hydrozoa, Mitrocomidae) and *Foersteria* Wehner, 1988 (Crustacea, Decapoda, Prosopidae), junior homonyms of *Foersteria* Szépligeti, 1896 (Insecta, Hymenoptera, Braconidae). Bulletin of the Mizunami Fossil Museum, **33**: 125-126.

COLLINS, J. S. H. & J. SAWARD, 2006. Three new genera and species of crabs from the Lower Eocene of London Clay of Essex, England. Bulletin of the Mizunami Fossil Museum, **33**: 67-76.

COLLINS, J. S. H., B. P. SCHULZ & S. L. JAKOBSEN, 2005. First record of brachyuran decapods (Crustacea, Decapoda) from Fur Formation (early Eocene) of Mors and Fur Island, Denmark. Bulletin of the Mizunami Fossil Museum, **32**: 17-22.

COLLINS, J. S. H. & R. SMITH, 1993. Ypresian (Lower Eocene) crabs (Decapoda, Crustacea) from Belgium. Bulletin de l'Institut Royal des Sciences Naturelles de Belgique, (Sciences de la Terre) **63**: 261-270, pls. 1, 2.

COLLINS, J. S. H. & A. WIERZBOWSKI, 1985. Crabs from the Oxfordian sponge megafacies of Poland. Acta Geologica Polonica, **35** (1-2): 73-88.

COLLINS, J. S. H. & R. J. WILLIAMS, 2004. A new genus and species of necrocarcinid crab (Crustacea, Brachyura) from the Upper Cretaceous of England. Bulletin of the Mizunami Fossil Museum, **31**: 33-35.

COLOSI, G., 1921. Un nuovo crostaceo fossile: "*Heteroglyphea paronae*". Atti Reale Accademia delle Scienze di Torino, **56**: 79-82.

— —, 1923. Una specie fossile de Gerionide (Decapodi Brachiuri). Bolettino della Società dei Naturalisti in Napoli, **35** [(2) **15**]: 248-255.

CONKLE, L. N., C. E. SCHWEITZER, R. M. FELDMANN & R. B. BLODGETT, 2006. A new species of *Tymolus* and a report on *Metacarcinus* from the Miocene of Alaska. Bulletin of the Mizunami Fossil Museum, **33**: 47-57.

COPE, E. D., 1871. On three extinct Astaci from the freshwater Tertiary of Idaho. Proceedings of the Ametican Philosophical Society, **11**: 605-607.

COPELAND, M. J., 1960. *Erymastacus bordenensis* a new Mesozoic decapod from the Canadian Arctic. Geological Survey of Canada Bulletin, **60**: 55-57.

COUCH, R. Q., 1851. Notice of a crustacean, new to Cornwall. Transactions of the Natural History and Antiquarian Society of Penzance, **2**: 13-14.

CRANE, M. D., 1981. Hexapod crabs of the genus *Goniocypoda* H. Woodward from the Upper Eocene of Hampshire. Zoological Journal of the Linnaean Society, **72**: 1-19.

CRANE, M. D. & J. QUAYLE, 1986. Two new hexapod crabs of the genus *Goniocypoda* Woodward (Crustacea, Decapoda) from the Hampshire Basin. Tertiary Research, **7**: 101-105.

CRAWFORD, R. S., 2008. A new species of fossil homolid crab (Decapoda: Brachyura) from the Río Foyel Formation (Paleogene), Río Negro Province, Argentina. Journal of Paleontology, **82**: 835-841.

CRAWFORD, R. S., R. M. FELDMANN, D. A. WAUGH, B. M. KELLEY & J. G. ALLEN, 2006. Decapod crustaceans from the Maastrichtian Fox Hills Formation. Bulletin of the Peabody Museum of Natural History, **47**: 3-28.

CREMA, C., 1895. Sopra alcuni decapodi terziarii del Piemonte. Atti Reale Accademia delle Scienze di Torino, **30**: 664-681, pl. 1.

CRESPIN, I., 1947. A fossil crab from the Lakes Entrance Oil Shaft, Gippsland, Victoria. Proceedings of the Royal Society of Victoria, **59**: 20-22.

CRESSWELL, A. W., 1886. Notes on some fossil crabs from the Miocene rocks of Corio Bay. Victorian Naturalist, **3** (6): 86-91.

CRÔNIER, C. & P. COURVILLE, 2004. A rich and highly endemic decapod crustacean fauna from the Middle Jurassic of north-east France. Palaeontology, **47** (4): 999-1014.

DALLONI, M., 1930. Étude géologique des Pyrénées Catalanes. Annales de la Faculté des Sciences de Marseille, **26**: 1-373.

DAMBORENEA, S. E. & M. O. MANCENIDO, 1987. Primer Glypheidae (Crustacea, Decapoda) de America del Sur en el Toarciano de la Provincia de Mendoza, Argentina. Notas del Museo de la Plata, **21** (106): 49-65.

DAMES, W., 1886. Ueber einige Crustaceen aus den Kreidablagerungen des Libanon. Zeitschrift der Deutschen Geologischen Gesellschaft, Berlin, **38**: 551-575.

DANA, J. D., 1849. Geology. United States Exploring Expedition during the years 1838-1842, under the Command of Charles Wilkes, U.S.N., **10**: 1-756, pls. 1-21. (C. Sherman, Philadelphia).

— —, 1851a. Paguridae. Conspectus Crustaceorum quae in Orbis Terrarum circumnavigatione, Carolo Wilkes e Classe Reipublicae Foederatae duce, lexit et descripsit. Proceedings of the Academy of Natural Sciences of Philadelphia, **5**: 267-272.

— —, 1851b. On the classification of the Cancroidea. American Journal of Science and Arts, (2) **12** (34): 121-131.

— —, 1851c. Conspectus Crustaceorum quae in Orbis Terrarum circumnavigatione, Carolo Wilkes e Classe Reipublicae Foederatae duce, lexit et descripsit. Pars VI. American Journal of Science and Arts, (2) **11**: 268-274.

— —, 1851d. On the classification of the maioid Crustacea or Oxyrhyncha. American Journal of Science and Arts, (2) **11**: 425-434.

— —, 1851e. On a new genus of Crustacea in the collection of the U.S. Exploring Expedition. American Journal of Science and Arts, (2) **11**: 223-224.

— —, 1851f. Crustacea Grapsoidea, (Cyclometopa, Edwardsii). Conspectus Crustaceorum quae in Orbis Terrarum circumnavigatione, Carolo Wilkes e Classe Reipublicae Foederatae duce, lexit et descripsit, 8. Proceedings of the Academy of Natural Sciences of Philadelphia, **5**: 247-254.

— —, 1852a. Parts I and II, Crustacea. U.S. Exploring Expedition during the Years 1838, 1839, 1840, 1841, 1842, under the Command of Charles Wilkes, U.S.N., **13**: 1-1618, 1 map; separate folio atlas with 96 pls. (C. Sherman, Philadelphia).

— —, 1852b. Conspectus Crustaceorum quae in Orbis Terrarum circumnavigatione, Carolo Wilkes e classe Reipublicae Foederatae duce, lexit et descripsit. Proceedings of the Academy of Natural Sciences in Philadelphia, **6**: 10-28.

— —, 1852c. Crustacea Cancroidea Corystoidea. Conspectus Crustaceorum quae in Orbis Terrarum circumnavigatione, Carolo Wilkes e classe Reipublicae Foederatae duce, lexit et descripsit. Proceedings of the Academy of Natural Sciences in Philadelphia, **6**: 73-86.

— —, 1853. On the classification and geographical distribution of Crustacea from the Report on Crustacea of the United States Exploring Expedition, under Captain Charles Wilkes, U.S.N., during the years 1838-1842: 1395-1592, 1 map. (C. Sherman, Philadelphia).

— —, 1854. Description of a new species of *Cryptopodia* from California. American Journal of Science and Arts, **18** (54): 430.

D'ARCHIAC, A., 1850. Description des fossiles du groupe nummulitique recueillis par M. S. P. Pratt et M. J. Delbos aus environs de Bayonne et de Dax. Mémoires de la Societé Géologique de France, (2) **2**: 397-456.

DAVIDSON, E., 1963. New linuparid crustaceans from the Upper Cretaceous of Texas. Bulletins of American Paleontology, **46** (206): 69-77.

— —, 1966. A new Paleocene crab from Texas. Journal of Paleontology, **40**: 211-213.

DAVIE, P. J. F. & P. K. L. NG, 2007. Two new subfamilies of Varunidae (Crustacea: Brachyura), with description of two new genera. Raffles Bulletin of Zoology, (Suppl.) **16**: 257-272.

DE ANGELI, A. & C. BESCHIN, 1998. *Ceronnectes*, nuovo genere di Brachiuro (Crustacea, Decapoda) dell'Eocene di Ungheria e Italia. Lavori – Società Veneziana di Scienze Naturali, **23**: 87-91.

— — & — —, 1999. I crostacei Matutinae (Brachyura, Calappidae) dell'Eocene del Veneto (Italia settentrionale). Studi e Ricerche – Associazione Amici del Museo – Museo Civico "G. Zannato", Montecchio Maggiore (Vicenza), **1999**: 11-22.

— — & — —, 2000. Due nuove specie di *Eopalicus* (Decapoda, Palicidae) nel Terziario del Veneto (Italia settentrionale). Studi e Ricerche – Associazione Amici del Museo – Museo Civico "G. Zannato", Montecchio Maggiore (Vicenza), **2000**: 7-12.

— — & — —, 2002. *Branchioplax albertii*, nuova specie di Goneplacidae (Crustacea, Decapoda) dell'Eocene di cava "Main" di Arzignano (Vicenza – Italia settentrionale). Lavori – Società Veneziana di Scienze Naturali, **27**: 125-130.

— — & — —, 2004. *Nucilobus bericus* sp. nov., nuovo crostaceo Leucosiidae dell'Eocene superiore del Veneto (Vicenza – Italia settentrionale). Lavori – Società Veneziana di Scienze Naturali, **29**: 119-122.

— — & — —, 2007. I crostacei Notopodinae (Brachyura, Raninidae) del Terziario del Vicentino. Studi e Ricerche – Associazione Amici del Museo – Museo Civico "G. Zannato", Montecchio Maggiore (Vicenza), **14**: 25-42.

— — & — —, 2008. Crostacei decapodi dell'Oligocene di Soghe e Valmarana (Monte Berici, Vicenza – Italia settentrionale). Studi e Ricerche – Associazione Amici del Museo – Museo Civico "G. Zannato", Montecchio Maggiore (Vicenza), **15**: 15-39.

DE ANGELI, A., C. BESCHIN & A. CHECCHI, 2005. Una nuova specie di Albuneidae Stimpson, 1858 dell'Eocene della Valle del Chiampo (Vicenza, NE Italia) e considerazioni sulle altre forme note (Decapoda, Anomura, Hippoidea). Lavori – Società Veneziana di Scienze Naturali, **30**: 85-91.

DE ANGELI, A. & A. GARASSINO, 2002. Galatheid, chirostylid and porcellanid decapods (Crustacea, Decapoda, Anomura) from the Eocene and Oligocene of Vicenza (N Italy). Memorie Società italiana di Scienze naturali e del Museo civico di Storia naturale in Milano, **30** (3): 1-40.

— — & — —, 2003. *Bittnerilia*, new genus for *Lambrus eocaenus* Bittner, 1883 (Decapoda, Brachyura, Calappidae) from the Middle Eocene of Veneto (N Italy). Atti della Società italiana di Scienze naturali e del Museo civico di Storia naturale in Milano, **144** (1): 13-22.

— — & — —, 2006a. New reports of decapod crustaceans from the Mesozoic and Cenozoic of Friuli-Venezia Giulia (NE Italy). Atti della Società italiana di Scienze naturali e del Museo civico di Storia naturale in Milano, **147**: 267-294.

— — & — —, 2006b. *Asthenognathus laverdensis* n. sp. (Decapoda: Brachyura: Pinnotheridae) from the Lower Oligocene of Laverda (Vicenza, NE Italy). Atti della Società italiana di Scienze naturali e del Museo civico di Storia naturale in Milano, **147** (2): 295-304.

— — & — —, 2008. *Pseudosquilla lessinea* n. sp. (Crustacea, Stomatopoda, Pseudosquillidae) and *Scyllarides bolcensis* n. sp. (Crustacea, Decapoda, Scyllaridae) from the Lower Eocene (Ypresian) of Monte Postale (Altissimo, Vicenza, NE Italy). Atti della Società italiana di Scienze naturali e del Museo civico di Storia naturale in Milano, **149**: 167-178.

DE ANGELI, A., A. GARASSINO & G. PASINI, 2009. New reports of anomurans and brachyurans from the Cenozoic of Tuscany (Italy). Atti della Società italiana di Scienze naturali e del Museo civico di Storia naturale in Milano, **150**: 163-196.

DE ANGELI, A. & S. MARANGON, 2001. *Paralbunea galantensis*, nuova specie di anomuro oligocenico del Bacino Ligure-Piemontese (Italia settentrionale). Studi Trentini di Scienze Naturali, (Acta Geologica) **76** [1999]: 99-105.

DE ANGELI, A. & L. MARCHIORI, 2009. *Eomatuta granosa* n. gen., n. sp. (Decapoda, Brachyura, Matutidae), nuovo crostaceo dell'Eocene dei Monti Berici (Vicenza, Italia settentrionale). Lavori – Società Veneziana di Scienze Naturali, **34**: 105-110.

DE ANGELI, A. & V. MESSINA, 1992. *Upogebia perarolensis* nuova specie di crostaceo del Terziario del Veneto (Italia). Lavori – Società Veneziana di Scienze Naturali, **17**: 183-191.

DE BERVILLE, P., 1857. Note sur une nouvelle espèce de Crustacé fossile trouvé dans le Calcaire Grossier Inférieur. Bulletin de la Societé Géologique de France, (2) **14**: 108-117, pl. 2.

DE BOUILLÉ, R., 1873. Paléontologie de Biarritz. Compte Rendu Travaux Congrès Scientifique de France, 39e session à Pau: 11, pl. 4.

— —, 1876. Paléontologie de Biarritz et des autres localités des Basses Pyrénées. (Pau).

DE FERRY, M., 1862. Note sur l'étage Bajocien des environs de Mâcon (Saône et Loire). Mémoires de la Société Linnéenne de Normandie, **12** [1860-61]: 1-46.

— —, 1865. Note sur les Crustacés de la base de l'étage bathonien de Mâcon (Saône et Loire). Bulletin de la Societé Linnéenne de Normandie, **9**: 365-375.

DE GRAVE, S. & K. MOOSA, 2004. A new species of the enigmatic shrimp genus *Pseudocheles* (Decapoda: Bresiliidae) from Sulawesi (Indonesia), with the designation of a new family Pseudochelidae. Crustacean Research, **33**: 1-9.

DE GREGORIO, A., 1884. Nuovi Decapodi titonici. Il Naturalista Siciliano, **3**: 134.

— —, 1895. Description des faunes tertiaires de la Vénétie. Note sur certain Crustacés (Brachioures) Éocéniques (avec un catalogue de tous les Crustacés de la Vénétie cités par les auteurs). Annales de Géologie et de Paléontologie, Palermo, **18**: 1-22, pls. 1-5.

DE HAAN, W., 1833-1850. Crustacea. In: P. F. VON SIEBOLD (ed.), Fauna Japonica sive Descriptio Animalium, quae in Itinere per Japoniam, Jussu et Auspiciis Superiorum, qui summum in India Batava Imperium Tenent, Suscepto, Annis 1823-1830 Collegit, Notis, Observationibus et Adumbrationibus Illustravit: i-xvii, i-xxxi, ix-xvi, 1-243, pls. A-J, L-Q, 1-55, circ. table 2. (J. Müller et Co., Lugduni Batavorum [= Leyden]).

DELL, R. K., 1969. A new Pliocene fossil crab of the genus (*Trichopeltarion*) from New Zealand. Records of the Canterbury Museum, **8**: 367-371.

DELLE CAVE, L., 1988. *Monodaeus bortolottii*, a new species of Brachyura (Crustacea, Decapoda) from the Pliocen of Tuscany (Italy). Bollettino della Società Paleontologica Italiana, **27** (2): 123-127, pls. 1, 2.

DE MAN, J. G., 1881. Carcinologica studies in the Leyden Museum. Notes of the Leyden Museum, **3**: 150-169.

— —, 1888. Bericht über die von Herrn Dr. J. Brock im indischen Archipel gesammelten Decapoden und Stomatopoden. Archiv für Naturgeschichte, **53** (1): 215-600, pls. 7-22.

— —, 1889. Über einige neue oder seltene indo-pacifische Brachyuren. Zoologische Jahrbücher, Iena, (Systematik) **4**: 409-452, pls. 9, 10.

— —, 1895. Bericht über die von Herrn Schiffscapitän Storm zu Atjeh, an den westlichen Küsten von Malakka, Borneo und Celebes sowie in der Java-See gesammelten Decapoden und Stomatopoden (Part 1). (Brachyura - *Doclea* zu *Cleistostoma*). Zoologische Jahrbücher, (Systematik, Ökologie, und Geographie der Tiere) **8**: 485-609, figs. 1-15.

— —, 1904. Beschreibung einiger Brachyuren Krebse aus post-tertiären Schichten der Minahassa, Celebes. Sammlungen des Geologischen Reichsmuseums in Leiden, (1) **7**: 254-277.

DERANIYAGALA, P. E. P., 1958. The Pleistocene of Ceylon: i-vi, 1-164, 58 pls. (Colombo).

DE SAINT LAURENT, M., 1973. Sur la systématique et la phylogénie des Thalassinidea: définition des familles des Callianassidae et des Upogebiidae et diagnose de cinq genres nouveaux (Crustacea Decapoda). Comptes Rendus Hebdomadaires des Séances de l'Académie des Sciences, Paris, (D) **277**: 513-516.

— —, 1979. Sur la classification et la phylogénie des Thalassinides: définition de la superfamille des Axioidea, de la sous-famille des Thalassininae et deux genres nouveaux (Crustacea Decapoda). Comptes Rendus Hebdomadaires des Séances de l'Académie des Sciences, Paris, (D) **288**: 1395-1397.

— —, 1980. Sur la classification et la phylogénie des Crustacés Décapodes Brachyoures. I. Podotremata Guinot, 1977, et Eubrachyura sect. nov. Comptes Rendus Hebdomadaires des Séances de l'Académie des Sciences, Paris, (D) **290**: 1265-1268.

— —, 1985. Remarques sur la distribution des Crustacés Décapodes. In: L. LAUBIER & C. I. MONNIOT (eds.), Peuplements profonds du golfe de Gascogne: 469-478. (IFREMER).

— —, 1988. Enoplometopidae, nouvelle superfamille de Crustacés Décapodes Astacidae. Comptes Rendus Hebdomadaires des Séances de l'Académie des Sciences, Paris, (Zoologie) **307**: 59-62.

DE SAUSSURE, H., 1853. Description de quelques Crustacés nouveaux de la côte occidentale du Mexique. Revue et Magasin de Zoologie, (2) **5**: 354-368, pls. 12, 13.

DESMAREST, A. G., 1822. Histoire naturelle des Crustacés fossiles. Les Crustacés proprement dits: 67-154, pls. 5-11. (F. G. Levrault, Paris).

— —, 1823. Crustacés Malacostracés. In: Dictionnaire des Sciences Naturelles, **28**: 138-425. (F. G. Levrault et Le Normant, Strasbourg and Paris).
— —, 1825. Considérations génerales sur la Classe des Crustacés: 1-446. (F. G. Levrault, Paris).
DE TRIBOLET, M., 1873. Recherches géologiques dans le Jura supérieur neuchâtelois vaudois. (Inaug.-Diss., Zürich).
— —, 1874. Description des Crustacés du terrain néocomien du Jura neuchâtelois et vaudois. Bulletin de la Société Géologique de France, (3) **2** [1873-1874]: 350-365, pl. 12.
— —, 1875. Description des Crustacés Décapodes des étages néocomien et urgonien de la Haute-Marne. Bulletin de la Société Géologique de France, (3) **3**: 451-459, pl. 15.
— —, 1876. Description de quelques espèces de Crustacés Décapodes du Valanginien, Néocomien et Urgonien de la Haute-Marne, du Jura et des Alpes. Bulletin de la Société des Sciences Naturelles de Neuchâtel, **10**: 868-870.
DI SALVO, G., 1933. I crostacei del Terziario Inferiore della Provincia di Palermo. Giornale di Scienze Naturale ed Echonomiche, **37** (3): 1-46, pls. 1, 2.
DOLLFUS, A., 1863. La faune kimmeridienne du Cap de la Hève. Essai d'une revision paléontologique. (Paris).
D'ORBIGNY, A., 1842. Voyage dans l'Amérique méridional, 1826-1833 (III). Géologie et Paléontologie: 1-561. (Paris).
DUMORTIER, E., 1864-1874. Études paléontologiques sur les dépots Jurassiques dans le bassin du Rhône, **1-4**.
DUPUY DE LÔME, E. & J. DE LA REVILLA, 1956. Dos especies fósiles nuevas de las provincias de Valencia y Alicante. Notas y Comunicaciones del Instituto Geológico y Minero de España, **43**: 5, fig. 1.
EBERT, T., 1887. Beitrag zur Kenntnis der tertiären Decapoden Deutschlands. Jahrbuch der Preussischen Geologischen Landesanstalt und Bergakademie in Berlin, **1886**: 262-271, pls. 8, 9.
EFFORD, I. E. & J. HAIG, 1968. Two new genera and three new species of crabs (Decapoda: Anomura: Albuneidae) from Australia. Australian Journal of Zoology, **16**: 897-914.
ÉTALLON, A., 1857. Esquisse d'une déscription géologique du Haut-Jura et en particulier des environs de St-Claude: 1-108. (J. B. Baillière et Fils, Paris).
— —, 1859. Description des Crustacés de la Haute Saône et du Haut-Jura. Bulletin de la Societé Géologique de France, (2) **16** [1858]: 169-205, pls. 3-6.
— —, 1861. Notes sur les Crustacés Jurassiques du bassin du Jura. Mémoires de la Societé de l'Agriculture, des Sciences et Lettres de la Haute Saône, **9**: 129-171, pl. 2.
ETHERIDGE, R., JR., 1914. The genus *Enoploclytia* in the Cretaceous rocks of Queensland. Records of the Australian Museum, **10**: 271-273.
— —, 1917. Description of some Queensland Paleozoic and Mesozoic fossils. 1. Queensland Lower Cretaceous Crustacea. Geological Survey of Queensland, Publication, **260**: 1-11, pls. 1, 2.
ETHERIDGE, R. E., JR. & A. R. MCCULLOCH, 1916. Sub-fossil Crustaceans from the coasts of Australia. Records of the Australian Museum, **11**: 1-14, pls. 1-7.
ETTER, W., 1994. A new penaeid shrimp (*Antrimpos mirigiolensis* n. sp., Crustacea, Decapoda) from the Middle Triassic of the Monte San Giorgio (Ticino, Switzerland). Neues Jahrbuch für Geologie und Paläontologie, Monatshefte, **1994**: 223-230.
EUDES-DESLONGCHAMPS, J. A., 1835. Mémoire pour servir à l'histoire naturelle des Crustacés fossils. Mémoire de la Societé Linnéenne de Normandie, **5**: 37-46, pl. 1.
— —, 1842. Mémoire pour servir à l'histoire naturelle des Crustacés fossiles. Mémoires de la Société Linnéenne de Normandie, Caen, **7**.

EYDOUX, F. & L. F. A. SOULEYET, 1842. Crustacés. In: Voyage autour du monde exécuté pendant les années 1836 et 1837 sur la corvette "La Bonite", commandée par M. Vaillant, etc., (Zoologie): 219-272, pls. 1-3. (Paris).

FABIANI, R., 1908. Palaeontologia dei Colli Berici. Memorie della Società Italiana di Scienze, (3) **15**: 45-248, pls. 1-6.

——, 1910. I crostacei terziari del Vicentino. Bolletin Museo Civico Vicenza, **1**: 1-40.

——, 1911. Di una nuova specie di *Phlyctenodes* del'Oligocene dei Berici. Bolletin Museo Civico Vicenza, **1** (3): 41-44.

——, 1915. Monografia sui terreni terziarii del Veneto, I. Il Paleogene. Memorie del'Istituto Geologico della Reale Università di Padova, **3**: 283-289.

FABRICIUS, J. C., 1775. Systema entomologiae, sistens insectorum classes, ordines, genera, species, adiectis synonymis, locis, descriptionibus, observationibus: 1-832.

——, 1777. Genera insectorum eorumque characteres naturales secundum numerum, figuram, situm et proportionem omnium partium orbis adjecta mantissa specierum nuper detectarum. (Bartsch, Chilonii).

——, 1787. Mantissa insectorvm sistens eorvm species nvper detectas adiectis characteribvs genericis, differentiis specificis, emendationibvs, observationibvs, **1**: i-xx, 1-348. (Proft, Hafniae [=Copenhagen]).

——, 1793. Entomologiae systematica emendata et aucta, secundum Classes, Ordines, Genera, Species, adjectis Synonimis, Locis, Observationibus, Descriptionibus: 1-519. (C. G. Proft et Storch, Hafniae [=Copenhagen]).

——, 1798. Supplementatione Entomologiae Systematicae: i-iv, 1-572. (C. G. Proft et Storch, Hafniae [=Copenhagen]).

FAXON, W., 1893. Preliminary descriptions of new species of Crustacea: Reports on the dredging operations off the west coast of Central America to the Galapagos, to the west coast of Mexico, and in the Gulf of California, in charge of Alexander Agassiz, carried on by the U.S. Fish Commission Steamer "Albatross", during 1891, Lieutenant Commander, Z. L. Tanner, U.S.N., Commanding, VI. Bulletin of the Museum of Comparative Zoology at Harvard College, **24** (7): 149-220.

FELDMANN, R. M., 1974. *Hoploparia riddlensis*, a new species of lobster (Decapoda: Nephropidae) from the Days Creek Formation (Hauterivian, Lower Cretaceous) of Oregon. Journal of Paleontology, **48** (3): 586-593.

——, 1979. *Eryma foersteri*, a new species of lobster (Decapoda) from the Jurassic (Callovian) of North America. American Museum Novitates, **2668**: 1-5.

——, 1984. *Haumuriaegla glaessneri* n. gen. and sp. (Decapoda; Anomura; Aeglidae) from Haumurian (Late Cretaceous) rocks near Cheviot, New Zealand. New Zealand Journal of Geology and Geophysics, **27**: 379-385.

——, 1989. *Metanephrops jenkinsi* n. sp. (Decapoda: Nephropidae) from the Cretaceous and Paleocene of Seymour Island, Antarctica. Journal of Paleontology, **63**: 64-69.

——, 1992. The genus *Lyreidus* De Haan, 1839 (Crustacea, Decapoda, Raninidae): systematics and biogeography. Journal of Paleontology, **66**: 943-957.

——, 1993. Additions to the fossil decapod crustacean fauna of New Zealand. New Zealand Journal of Geology and Geophysics, **36**: 201-211.

——, 1994. *Antarctomithrax thomsoni*, a new genus and species of crab (Brachyura; Majidae) from the La Meseta Formation of Seymour Island, Antarctica. Journal of Paleontology, **66** (1): 174-176.

——, 1998. *Paralomis debodeorum*, a new species of decapod Crustacean from the Miocene of New Zealand: first notice of the Lithodidae in the fossil record. New Zealand Journal of Geology and Geophysics, **41**: 35-38.

FELDMANN, R. M., E. E. B. AWOUTA & J. WELSHENBAUGH, 1976. *Necrocarcinus siouxensis*, a new species of calappid crabs (Crustacea: Decapoda) from the Fox Hills Formation (Cretaceous: Maastrichtian) of North Dakota. Journal of Paleontology, **50** (5): 985-990.

FELDMANN, R. M. & R. K. BEARLIN, 1988. *Linuparus korura* n. sp. (Decapoda: Palinura) from the Bortonian (Eocene) of New Zealand. Journal of Paleontology, **62** (2): 245-250.

FELDMANN, R. M., K. L. BICE, C. E. SCHWEITZER HOPKINS, E. W. SALVA & K. PICKFORD, 1998. Decapod crustaceans from the Eocene Castle Hayne Limestone, North Carolina: Paleoceanographic implications. The Paleontological Society Memoir, **48** (Supplement to Journal of Paleontology, **67**): 1-28.

FELDMANN, R. M., G. A. BISHOP & T. W. KAMMER, 1977. Macrurous decapods from the Bearpaw Shale (Cretaceous: Campanian) of north-eastern Montana. Journal of Paleontology, **51**: 1161-1180.

FELDMANN, R. M., S. CASADÍO, L. CHIRINO-GALVÉZ & M. AGUIRRE URRETA, 1995. Fossil decapod crustaceans from the Jagüel and Roca Formations (Maastrichtian-Danian) of the Neuquén Basin, Argentina. Paleontological Society Memoir, **43**: 1-22.

FELDMANN, R. M., L. CHIRINO-GALVEZ, G. L. MASON, J. L. ANDERSON, P. W. DUNCAN, R. A. WARD & D. R. SALEM, 1993. Fossil Penaeidae (Crustacea: Decapoda) from the Loyola Formation, Ecuador. Annals of Carnegie Museum, **62**: 229-243.

FELDMANN, R. M. & M. J. COPELAND, 1988. A new species of erymid lobster from Lower Jurassic strata (Sinemurian/Pliensbachian), Fernie Formation, southwestern Alberta. Contributions to Canadian Paleontology, Geological Survey of Canada, Bulletin, **379**: 93-101, pls. 4.1, 4.2.

FELDMANN, R. M. & J. A. CRAME, 1998. The significance of a new nephropid lobster from the Miocene of Antarctica. Palaeontology, **41**: 807-814.

FELDMANN, R. M., G. CRISP & D. PIRRIE, 2002. A new species of glypheoid lobster, *Pseudoglyphaea foersteri* (Decapoda: Astacidea: Mecochiridae) from the Lower Jurassic (Pliensbachian) of Raasay, Inner Hebrides, U.K. Palaeontology, **45** (1): 23-32.

FELDMANN, R. M., M. FERNANDA RODRIGUEZ, G. A. MARTINEZ & M. AGUIRRE-URRETA, 1997. *Costacopluma salamanca* new species (Decapoda, Retroplumidae) from the Salamanca Formation (Danian) of Patagonia, Argentina. Journal of Paleontology, **71**: 125-130.

FELDMANN, R. M. & R. E. FORDYCE, 1996. A new cancrid crab from New Zealand. New Zealand Journal of Geology and Geophysics, **39**: 509-513.

FELDMANN, R. M. & A. GAŹDICKI, 1997. A new species of *Glyphea* (Decapoda: Palinura) from the La Meseta Formation (Eocene) of Seymour Island, Antarctica. Acta Palaeontologica Polonica, **42**: 437-445.

FELDMANN, R. M., L. GRANDE, C. P. BIRKHEIMER, J. T. HANNIBAL & D. L. MCCOY, 1981. Decapod fauna of the Green River Formation (Eocene) of Wyoming. Journal of Paleontology, **55** (4): 788-799.

FELDMANN, R. M. & J. W. HAGGART, 2007. A new species of lobster (Astacidea, Erymidae) from the Smithers Formation (Middle Jurassic) of British Columbia, Canada. Canadian Journal of Earth Sciences, **44**: 1791-1796.

FELDMANN, R. M. & F. D. HOLLAND, JR., 1971. Lobsters from the Cannonball Formation (Paleocene) of North Dakota. Journal of Paleontology, **45**: 838-843.

FELDMANN, R. M., A. KOLAHDOUZ, B. BIRANVAND & G. SCHWEIGERT, 2007. A new family, genus and species of lobster (Decapoda: Achelata) from the Gadvan Formation (Early Cretaceous) of Iran. Journal of Paleontology, **81**: 405-407.

FELDMANN, R. M., I. LAZĂR & C. E. SCHWEITZER, 2006. New crabs (Decapoda: Brachyura: Prosopidae) from Jurassic (Oxfordian) sponge bioherms of Dobrogea, Romania. Bulletin of the Mizunami Fossil Museum, **33**: 1-20.

FELDMANN, R. M., R.-Y. LI & C. E. SCHWEITZER, 2007. A new family, genus, and species of crab (Crustacea, Decapoda, Brachyura) from the Upper Cretaceous (Campanian) of Manitoba, Canada. Canadian Journal of Earth Science, **44**: 1741-1752.

FELDMANN, R. M. & R. G. MARTINS-NETO, 1995. *Costacopluma nordestina* n. sp. (Decapoda: Retroplumidae) from the Maria Farinha Formation (Paleocene) of Brazil. Journal of Paleontology, **69**: 610-611.

FELDMANN, R. M. & M. MAXEY, 1980. *Raninella carlilensis*, a new raninid crab from the Carlile Shale (Turonian) of Kansas. Journal of Paleontology, **54** (4): 858-861.

FELDMANN, R. M. & P. A. MAXWELL, 1990. Late Eocene decapod Crustacea from North Westland, South Island, New Zealand. Journal of Paleontology, **64**: 779-797.

— — & — —, 1999. A new species of glypheid lobster, *Glyphea christeyi* (Decapoda, Palinura), from the Eocene (Bortonian) Waihao Greensand, South Canterbury, New Zealand. New Zealand Journal of Geology and Geophysics, **42**: 75-78.

FELDMANN, R. M. & C. B. MCPHERSON, 1980. Fossil decapod crustaceans of Canada. Geological Survey of Canada Paper, **79-16**: 1-20.

FELDMANN, R. M., P. M. O'CONNOR, N. J. STEVENS, M. D. GOTTFRIED, E. M. ROBERTS, S. NGASALA, E. L. RASMUSSON & S. KAPILIMA, 2007. A new freshwater crab (Decapoda: Brachyura: Potamonautidae) from the Paleogene of Tanzania, Africa. Neues Jahrbuch für Geologie und Paläontologie, Abhandlungen, **244**: 71-78.

FELDMANN, R. M. & M. POLE, 1994. A new species of *Paranephrops* White, 1842: a fossil freshwater crayfish (Decapoda: Parastacidae) from the Manuherikia Group (Miocene), central Otago, New Zealand. New Zealand Journal of Geology and Geophysics, **37**: 163-167.

FELDMANN, R. M. & R. W. PORTELL, 2007. First report of *Costacopluma* Collins & Morris, 1975 (Decapoda: Brachyura: Retroplumidae) from the Eocene of Alabama, U.S.A. Journal of Crustacean Biology, **27**: 90-96.

FELDMANN, R. M. & M. DE SAINT LAURENT, 2002. *Glyphaea foresti* n. sp. (Decapoda) from the Cenomanian of Northern Territory, Australia. Crustaceana, **75** (3-4): 359-373.

FELDMANN, R. M. & C. E. SCHWEITZER, 2009. Revision of Jurassic Homoloidea De Haan, 1839, from the Ernstbrunn and Štramberk limestones, Austria and the Czech Republic. Annalen des Naturhistorischen Museums in Wien, (A) **111**: 183-206.

FELDMANN, R. M., C. E. SCHWEITZER & A. ENCINAS, 2005. New decapods from the Navidad Formation (Miocene) of Chile. Journal of Crustacean Biology, **25**: 427-449.

FELDMANN, R. M., C. E. SCHWEITZER & R. M. GREEN, 2008. Unusual Albian (Early Cretaceous) Brachyura (Homoloidea: Componocancroidea new superfamily) from Montana and Wyoming, U.S.A. Journal of Crustacean Biology, **28**: 502-509.

FELDMANN, R. M., C. E. SCHWEITZER, P. A. MAXWELL & B. M. KELLEY, 2008. Fossil isopod and decapod crustaceans from the Kowai Formation (Pliocene) near Makikihi, South Canterbury, New Zealand. New Zealand Journal of Geology and Geophysics, **51**: 43-58.

FELDMANN, R. M., C. E. SCHWEITZER & D. MCLAUGHLIN, 2006. Additions to the records for decapod Crustacea from Motunau and Glenafric beaches, North Canterbury, New Zealand. New Zealand Journal of Geology and Geophysics, **49**: 417-427.

FELDMANN, R. M., C. E. SCHWEITZER, C. M. REDMAN, N. J. MORRIS & D. J. WARD, 2007. New Cretaceous lobsters from the Kyzylkum Desert of Uzbekistan. Journal of Paleontology, **81**: 701-713.

FELDMANN, R. M., C. E. SCHWEITZER & W. WAHL, 2008. New species of Glaessneropsidae from Western Interior. Journal of Paleontology, **82**: 1030-1034.

FELDMANN, R. M. & A. L. TITUS, 2006. *Eryma jungostrix*, n. sp. (Decapoda; Erymidae) from the Redwater Shale Member of the Stump Formation (Jurassic; Oxfordian) of Utah. Journal of Paleontology, **26**: 63-68.

FELDMANN, R. M. & D. M. TSHUDY, 1988. Macruran decapods, and their epibionts, from the López de Bertodano Formation (Upper Cretaceous), Seymour Island, Antarctica. In: R. M. FELDMANN & M. O. WOODBURNE (eds.), Geology and paleontology of Seymour Island, Antarctic Peninsula. Geological Society of America Memoir, **169**: 291-301, (Boulder, Colorado).

FELDMANN, R. M., D. M. TSHUDY & M. R. A. THOMSON, 1993. Late Cretaceous and Paleocene decapod crustaceans from James Ross Basin, Antarctic Peninsula. The Paleontological Society Memoir, **28**: i-iv, 1-41.

FELDMANN, R. M., F. J. VEGA, S. P. APPLEGATE & G. A. BISHOP, 1998. Early Cretaceous arthropods from the Tlayúa Formation at Tepexi de Rodríguez, Puebla, Mexico. Journal of Paleontology, **72** (1): 79-90.

FELDMANN, R. M., F. J. VEGA, P. GARCÍA-BARRERA, R. RICO-MONTIEL & L. MARTÍNEZ LÓPEZ, 1995. A new species of *Meyeria* (Decapoda: Mecochiridae) from the San Juan Raya Formation (Aptian: Cretaceous), Puebla State, Mexico. Journal of Paleontology, **69**: 402-406.

FELDMANN, R. M., F. J. VEGA, L. MARTÍNEZ-LÓPEZ, K. A. GONZÁLEZ-RODRÍGUEZ, O. GONZÁLEZ-LEÓN & M. R. FERNÁNDEZ-BARAJAS, 2007. Crustacea from the Muhi Quarry (Albian – Cenomanian), and a review of two Aptian Mecochiridae from Mexico. Annals of Carnegie Museum, **76** (3): 135-144.

FELDMANN, R. M., F. J. VEGA, A. B. TUCKER, P. GARCÍA-BARRERA & J. AVENDAÑO, 1996. The oldest record of *Lophoranina* (Decapoda: Raninidae) from the Late Cretaceous of Chiapas, southeastern Mexico. Journal of Paleontology, **70** (2): 296-303.

FELDMANN, R. M. & T. VILLAMIL, 2002. A new carcineretid crab (Upper Turonian, Cretaceous) of Colombia. Journal of Paleontology, **76** (4): 718-724.

FELDMANN, R. M., T. VILLAMIL & E. G. KAUFFMAN, 1999. Decapod and stomatopod crustaceans from mass mortality Lagerstatten: Turonian (Cretaceous) of Colombia. Journal of Paleontology, **73**: 91-101.

FELDMANN, R. M. & R. R. WEST, 1978. *Huhatanka*, a new genus of lobster (Decapoda: Mecochiridae) from the Kiowa Formation (Cretaceous: Albian) of Kansas. Journal of Paleontology, **52** (66): 1219-1226, pl. 1.

FELDMANN, R. M. & M. T. WILSON, 1988. Eocene decapod crustaceans from Antarctica. In: R. M. FELDMANN & M. O. WOODBURNE (eds.), Geology and paleontology of Seymour Island, Antarctic Peninsula. Geological Society of America, Memoir, **169**: 465-488.

FELDMANN, R. M. & W. J. ZINSMEISTER, 1984a. First occurrence of fossil decapod crustaceans (Callianassidae) from the McMurdo Sound region, Antarctica. Journal of Paleontology, **58**: 1041-1045.

— — & — —, 1984b. New fossil crabs (Decapoda: Brachyura) from the La Meseta Formation (Eocene) of Antarctica: paleogeographic and biogeographic implications. Journal of Paleontology, **58**: 1046-1061.

FENGLIN, L., 1975. On Triassic crayfish – *Sinopemphix* gen. nov. Acta Geologica Sinica, **2**: 136-141, pl. 1. [In Chinese.]

FERRY, M. DE *see* DE FERRY, M.

FISCHER-BENZON, R. VON *see* VON FISCHER-BENZON, R.

FLEMING, C. A., 1981. A new grapsid crab from the upper Miocene of New Zealand. Journal of the Royal Society of New Zealand, **11**: 103-108.

FONTANNES, F., 1884. Note sur quelques gisements nouveaux des terrains miocènes du Portugal et description d'un Portunien du genre *Achelous*. Annales des Sciences Géologique, Paris, **16**: [article 3], pls. 6, 7.

— —, 1885. Nouvelles contributions à la faune et à la flore des marnes pliocènes a *Brissopsis* d'Eurre (Drôme). (Lyon).

FOREST, J., 1995. Crustacea Decapoda Anomura: présence du genre *Ciliopagurus* Forest, 1995 (Diogenidae) au Badénien (Miocène moyen): *C. substriatiformis* (Lőrenthey, 1929). In: A. CROSNIER (ed.), Résultats des campagnes MUSORSTOM, **13**. Mémoires du Muséum National d'Histoire Naturelle, Paris, **163**: 151-154.

FORIR, H., 1887. Contributions à l'étude du système Crétacé de la Belgique. II. Études complémentaires sur les Crustacés. Annales de la Societé Géologique de Belgique, **14**: 155-175.

— —, 1889. Contributions à l'étude du système Crétacé de la Belgique. IV. Troisième note sur les Poissons et Crustacés nouveaux ou peu connues. Annales de la Societé Géologique de Belgique, **16**: 445-460, pl. 74.

FORSKÅL, P., 1775. Descriptiones animalium, avium, amphibiorum, piscium, insectorum, vermium; quae in itinere orientali observavit Petrus Forskål: 1-164. (Mölleri, Hauniae [= Copenhagen]).

FÖRSTER, R., 1966. Uber die Erymiden, eine alte konservative Familie der mesozoischen Dekapoden. Palaeontographica, **125**: 61-175.

— —, 1967a. Zur Kenntnis natanter Jura-Dekapoden. Mitteilungen der Bayerischen Staatssammlung für Paläontologie und Historische Geologie, **7**: 157-174, pl. 8.

— —, 1967b. Die reptanten Dekapoden der Trias. Neues Jahrbuch für Geologie und Paläontologie, Abhandlungen, **128** (2): 136-194.

— —, 1968. *Paranecrocarcinus libanoticus* n. sp. (Decapoda) und die Entwicklung der Calappidae in der Kreide. Mitteilungen der Bayerischen Staatssammlung für Paläontologie und Historische Geologie, **8**: 167-195.

— —, 1970a. Neue Dekapoden-Reste aus der Oberkreide von Moçambique, Norddeutschland und den bayerischen Alpen. Paläontologische Zeitschrift, **44**: 134-144.

— —, 1970b. Zwei neue brachyure Krebse aus dem Paläozän des Haunsberges nördlich von Salzburg. Mitteilungen der Bayerischen Staatssammlung für Paläontologie und Historische Geologie, **10**: 241-251, pl. 4.

— —, 1971. The Mecochiridae, a specialized family of the Mesozoic Glypheoidea (Crustacea, Decapoda). Neues Jahrbuch für Geologie und Paläontologie, Abhandlungen, **137**: 396-421.

— —, 1979. Decapod crustaceans from the Middle Miocene (Badenian) deposits of southern Poland. Acta Geologica Polonica, **29**: 89-106.

— —, 1980. Krebsfunde aus dem Obersten Lias und dem Untersten Dogger von Mistelgau bei Bayreuth, Oberfranken. Geologische Blätter für Nordost-Bayern und angrenzende Gebiete, **30** (2): 73-90.

— —, 1984. Bärenkrebse (Crustacea, Decapoda) aus dem Cenoman des Libanon und dem Eozän Italiens. Mitteilungen der Bayerischen Staatssammlung für Paläontologie und Historische Geologie, **24**: 57-66.

— —, 1985. Frühe Anomuren und Brachyuren (Decapoda, Crustacea) aus dem Mittleren Dogger. Mitteilungen der Bayerischen Staatssammlung für Paläontologie und Historische Geologie, **25**: 45-60.

— —, 1986. Der erste Nachweis eines brachyuren Krebses aus dem Lias (Oberes Pliensbach) Mitteleuropas. Mitteilungen der Bayerischen Staatssammlung für Paläontologie und Historische Geologie, **26**: 25-31.

FÖRSTER, R. & G. CHONG, 1976. *Chilenophoberus atacamensis*, a new decapod Crustacean from the Middle Oxfordian of the Cordillera de Domeyko, northern Chile. Neues Jahrbuch für Geologie und Paläontologie, Monatshefte, **1976**: 145-156.

FÖRSTER, R. & M. D. CRANE, 1984. A new species of the penaeid schrimp *Aeger* Münster (Crustacea, Decapoda) from the Upper Triassic of Somerset, England. Neues Jahrbuch für Geologie und Paläontologie, Monatshefte, **1984** (H): 455-462.

FÖRSTER, R., A. GAŹDZICKI & R. WRONA, 1985. First record of a homolodromiid crab from a Lower Miocene glacio-marine sequence of West Antarctica. Neues Jahrbuch für Geologie und Paläontologie, Monatshefte, **6**: 340-348.

FÖRSTER, R. & A. VON HILLEBRANDT, 1984. Das Kimmeridge des Profeta-Jura in Nordchile mit einer *Mecochirus-Favreina*-Vergesellschaftung (Crustacea, Decapoda-Ichnogenus). Mitteilungen der Bayerischen Staatssammlung für Paläontologie und Historische Geologie, **24**: 67-84.

FÖRSTER, R. & R. MUNDLOS, 1982. Krebse aus dem Alttertiär von Helmstedt und Handorf (Niedersachsen). Palaeontographica, (A) **179**: 148-184.

FÖRSTER, R. & H. RIEBER, 1982. Der älteste Vertreter der Gattung *Palaeastacus* (Crustacea, Decapoda), *Palaeastacus argoviensis* n. sp., aus dem Unteren Dogger der Nordschweiz. Eclogae Geologicae Helvetiae, **75** (3): 773-778.

FÖRSTER, R. & W. STINNESBECK, 1987. Zwei neue Krebse, *Callianasssa saetosa* n. sp. und *Homolopsis chilensis* n. sp. (Crustacea, Decapoda) aus der Oberkreide Zentral-Chiles. Mitteilungen der Bayerischen Staatssammlung für Paläontologie und Historische Geologie, **27**: 51-65.

FRAAIJE, R. H. B., 2002a. The first record of albuneid crabs (Crustacea, Decapoda) from the Cretaceous. Bulletin of the Mizunami Fossil Museum, **29**: 69-72. [See also R. H. B. FRAAYE.]

— —, 2002b. New calappid crabs (Crustacea, Decapoda, Brachyura) from the Late Maastrichtian of the Netherlands. Journal of Paleontology, **76**: 913-917.

FRAAIJE, R. H. B., B. W. M. VAN BAKEL & J. W. M. JAGT, 2007. A new species of *Goniocypoda* and the first record of *Glyphithyreus wetherelli* (Bell, 1858) (Decapoda, Brachyura) from the Eocene of Nieuwvliet-Bad, the Netherlands. In: A. GARASSINO, R. M. FELDMANN & G. TERUZZI (eds.), 3rd Symposium on Mesozoic and Cenozoic decapod crustaceans, Museo di Storia Naturale di Milano, May 23-25, 2007. Memorie della Società Italiana di Scienze Naturali e del Museo di Storia Naturale di Milano, **35** (2): 37-42.

— —, — — & — —, 2008. *Albunea turritellacola*, a new sand crab (Anomura, Albuneidae) from the Lower Miocene of southwest France. Bulletin of the Mizunami Fossil Museum, **34**: 17-22, 2 figs., 1 table, 1 pl.

FRAAIJE, R. H. B., B. W. M. VAN BAKEL, J. W. M. JAGT & P. ARTAL, 2008. New decapod crustaceans (Anomura, Brachyura) from Mid-Cretaceous reefal deposits at Monte Orobe (Navarra, northern Spain), and comments on related type-Maastrichtian material. Bulletin de l'Institut Royal des Sciences Naturelles de Belgique, (Sciences de la Terre) **78**: 193-208.

FRAAIJE, R. H. B., B. W. M. VAN BAKEL, J. W. M. JAGT & Y. COOLE, 2006. Two new Paleogene species of mud shrimp (Crustacea, Decapoda, Upogebiidae) from Europe and North America. Bulletin of the Mizunami Fossil Museum, **33**: 77-85.

FRAAIJE, R. H. B., B. W. M. VAN BAKEL, J. W. M. JAGT, A. A. KLOMPMAKER & P. ARTAL, 2009. A new hermit crab (Crustacea, Anomura, Paguroidea) from the Mid-Cretaceous of Navarra, northern Spain. Boletín de la Sociedad Geológica Mexicana, **61**: 13-16.

FRAAIJE, R. [H. B.], J. HANSEN & T. HANSEN, 2005. Late Miocene decapod faunas from Gram, Denmark. Palaeontos, **7**: 51-61.

FRAAIJE, R. H. B., F. J. VEGA, B. W. M. VAN BAKEL & L. M. GARIBAY-ROMERO, 2006. Late Cretaceous dwarf decapods from Guerrero, southern Mexico and their migration patterns. Contributions to Zoology, **75** (3/4): 121-132.

FRAAS, O., 1855. Beiträge zum obersten Weissen Jura in Schwaben. Jahreshefte des Vereins für Vaterländische Naturkunde in Württemberg, **11**: 76-106.

— —, 1867. Geologisches aus dem Orient. Jahreshefte des Vereins für Vaterländische Naturkunde in Württemberg, **23**: 145-362, pls. 4-6.

— —, 1878. Geologisches aus dem Libanon. Jahreshefte des Vereins für Vaterländische Naturkunde in Württemberg, **34**: 257-391, pls. 3-8.

FRAAYE, R. H. B., 1994. Early Paleocene crabs (Crustacea, Decapoda) from the Middle Vistula Valley, central Poland. Acta Geologica Polonica, **44** (3-4): 261-266, pls. 1, 2. [See also R. H. B. FRAAIJE.]

— —, 1995. A new raninid crab, *Pseudorogueus rangiferus* (Decapoda, Crustacea) from the Eocene of Spain. Estudios Geol., **51**: 65-67.

— —, 1996a. Two new crabs, *Graptocarcinus maastrichtiensis*, and *Caloxanthus kuypersi* (Crustacea, Decapoda), from the Maastrichtian of the Netherlands. Journal of Paleontology, **70** (3): 463-465.

— —, 1996b. A new Tethyan migrant: *Cretachlorodius enciensis* n. gen., n. sp. (Crustacea, Decapoda), from the Maastrichtian type area. Journal of Paleontology, **70**: 463-465.

— —, 1996c. Late Cretaceous swimming crabs: radiation, migration, competition, and extinction. Acta Geologica Polonica, **46**: 269-278.

FRAAYE, R. H. B. & B. W. M. VAN BAKEL, 1998. New raninid crabs (Crustacea, Decapoda, Brachyura) from the late Maastrichtian of the Netherlands. Geologie en Mijnbouw, **76**: 293-299.

FRAAYE, R. H. B. & J. S. H. COLLINS, 1987. A new crab, *Prehepatus werneri* (Crustacea: Decapoda) from the Maastrichtian of the Netherlands. Journal of Paleontology, **61**: 549-551.

— — & — —, 1996. Two new decapod crustaceans from the Portlandian of Dorset, England. Proceedings of the Geologists' Association, **107**: 323-326.

FRENTZEN, K., 1937a. *Mecochirus eckerti* nov. spec. Aus dem Lias Epsilon (Posidonienschiefer) von Langenbrüden. Beiträge zur Naturkundlichen Forschung in Südwestdeutschland, **2** (1): 103-105.

— —, 1937b. Über einige Decapoden aus dem Lias Delta Schmabens. Beiträge zur Naturkundlichen Forschung in Südwestdeutschland, **2** (1): 106-115.

FRITSCH, A., 1867. Über die *Callianassa* der böhmischen Kreideformation. Abhandlungen der Königlichen Böhmischen Gesellschaft der Wissenschaften, (5) **15**: 1-12.

— —, 1872. Ueber *Palaemon exul*, eine neue Crustacee aus dem Polirshiefer von Kutschlin bei Bilin in Böhmen. Sitzungsberichte der Königlichen Böhmischen Gesellschaft der Wissenschaften, **1872**: 37-38.

— —, 1893. Studien im Gebiete der böhmischen Kreideformation. Paläontologische Untersuchungen der einzelnen Schichten. V. Priesener Schichten. Archiv der Naturwissenschaftlichen Landesdurchforschung von Böhmen, **9** (1).

FRITSCH, A. & J. KAFKA, 1887. Die Crustaceen der böhmischen Kreideformation: 1-53, 10 pls. (Selbstverlag in Commission von F. Rivnác, Prag).

FRITSCH, K. VON *see* VON FRITSCH, K.

FUJIYAMA, I. & M. TAKEDA, 1980. A fossil raninid crab from the Poronai Formation, Hokkaido, Japan. In: H. IGO & H. NODA (eds.), Professor Saburo Kanno Memorial volume, the University of Tsukuba, Ibaraki: 339-342, pls. 39, 40.

GABB, W. M., 1864. Description of the Cretaceous fossils. Geologic Survey of California, Paleontology, (4) **1**: 55-236.

GALEOTTI, H., 1837. Mémoire sur la constitution géognostique de la province Brabant. Mémoires Couronnées par l'Académie Royal de Belgique: 12.

GALIL, B. S., 2001. A revision of *Myra* Leach, 1817 (Crustacea: Decapoda: Leucosioidea). Zoologische Mededelingen, Leiden, **75** (24): 409-446.

——, 2003a. Four new genera of leucosiid crabs (Crustacea: Brachyura: Leucosiidae) for three new species and nine species previously in the genus *Randallia* Stimpson, 1857, with a redescription of the type species, *R. ornata* (Randall, 1939). Proceedings of the Biological Society of Washington, **116** (2): 395-422.

——, 2003b. Contribution to the knowledge of Leucosiidae II. *Euclosia* gen. nov. (Crustacea: Brachyura). Zoologische Mededelingen, Leiden, **77** (20): 331-347.

——, 2005. Contribution to the knowledge of Leucosiidae III. *Urnalana* gen. nov. (Crustacea: Brachyura). Zoologische Mededelingen, Leiden, **79** (2): 9-40.

GALIL, B. S. & P. F. CLARK, 1994. A revision of the genus *Matuta* Weber, 1795 (Crustacea: Brachyura: Calappidae). Zoologische Verhandelingen, Leiden, **294**: 3-55, pls. 1-14.

GALIL, B. & M. TAKEDA, 1986. Resurrection of the genus *Jonesius* and the establishment of a new genus: commensal crabs associated with corals from the Indo-Pacific Ocean. Bulletin of the National Science Museum, Tokyo, (A) **12**: 163-171.

GARASSINO, A., 1994. The macruran decapod crustaceans of the Upper Cretaceous of Lebanon. Paleontologia Lombarda, Milano, (nuova serie) **3**: 1-27, pls. 1-13.

——, 1996. The family Erymidae Van Straelen, 1924 and the superfamily Glypheoidea Zittel, 1885 in the Sinemurian of Osteno in Lombardy (Crustacea, Decapoda). Atti della Società italiana di Scienze naturali e del Museo civico di Storia naturale in Milano, **135** (2): 333-373.

——, 1997. La specie *Glyphea tonelloi* n. sp. (Crustacea, Decapoda) nel Cretacico Inferiore (Barremiano Superiore-Aptiano) della Valle del Torrente Cornappo (Udinc, NE Italia). Gortania – Atti del Museo Friulano di Storia Naturale, **19**: 85-93.

——, 1998. Nuovo studio sui crostacei decapodi del Cretacico Inferiore (Barremiano Superiore-Aptiano) della Valle del Torrente Cornappo (Udine, NE Italia). Gortania – Atti del Museo Friulano di Storia Naturale, **20**: 59-73.

——, 2000. *Glyphea rigoi* n. sp. (Crustacea, Decapoda) della Dolomia di Forni Norico, (Triassico Superiore) della Carnia (Udine, NE Italia). Gortania – Atti del Museo Friulano di Storia Naturale, **22**: 59-64.

——, 2001. New decapod crustaceans in the Cenomanian (Upper Cretaceous) of Lebanon. Atti della Società italiana di Scienze naturali e del Museo civico di Storia naturale in Milano, **141** (2): 237-250.

GARASSINO, A., A. DE ANGELI, L. M. GALLO & G. PASINI, 2004. Brachyuran and anomuran fauna from the Cenozoic of Piedmont (NW Italy). Atti della Società italiana di Scienze naturale e del Museo civico di Storia naturale in Milano, **145** (2): 251-281.

GARASSINO, A., A. DE ANGELI & G. PASINI, 2008. New decapod assemblage from the Upper Cretaceous (Cenomanian-Turonian) of Gara Sbaa, southeastern Morocco. Atti della Società italiana di Scienze naturali e del Museo civico di Storia naturale in Milano, **149** (1): 37-367.

— —, — — & — —, 2009a. A new hermit crab (Crustacea, Anomura, Paguroidea) from the Late Cretaceous (Cenomanian) of Lebanon. Atti della Società italiana di Scienze naturali e del Museo civico di Storia naturale in Milano, **150** (2): 215-228.

— —, — — & — —, 2009b. *In situ* hermit crab (Crustacea, Anomura, Paguroidea) from the Early Eocene (Ypresian) of NE Italy. Atti della Società italiana di Scienze naturali e del Museo civico di Storia naturale in Milano, **150** (2): 229-238.

GARASSINO, A., P. ARTAL & G. PASINI, 2009a. *Jabaloya aragonensis* n. gen., n. sp. (Crustacea, Decapoda, Mecochiridae) and *Cedrillosia jurassica* n. gen., n. sp. (Crustacea, Decapoda, Glypheidae) from the Jurassic of Teruel Province (Aragon, Spain). Atti della Società italiana di Scienze naturali e del Museo civico di Storia naturale in Milano, **150** (1): 197-206.

— —, — — & — —, 2009b. New records of decapod macrurans from the Cretaceous of Catalonia and the Province of Castellón (Spain). Bulletin of the Mizunami Fossil Museum, **35**: 87-95.

— —, — — & — —, 2009c. *Upogebia miocenica* n. sp. (Crustacea, Thalassinidea, Upogebiidae) from the Miocene of Catalonia (Spain). Atti della Società italiana di Scienze naturali e del Museo civico di Storia naturale in Milano, **150** (1): 61-68.

— —, — — & — —, 2009d. New report of decapod macrurans from the Eocene of Catalonia and Aragón (Spain). Atti della Società italiana di Scienze naturali e del Museo civico di Storia naturale in Milano, **150** (2): 207-214.

GARASSINO, A. & S. BRAVI, 2003. *Palaemon antonellae* n. sp. (Crustacea, Decapoda, Caridea) from the the Lower Cretaceous "Platydolomite" of Profeti (Caserta, S Italy). Journal of Paleontology, **77** (3): 589-592.

GARASSINO, A. & G. BRETON, 2009. *Palinurus destombesi* n. sp. (Crustacea, Decapoda, Palinuridae) de l'Albien (Crétacé Inférieur) de Wissant (Pas-de-Calais, France). In: C. COLLETÉ (ed.), Stratotype Albine. (Collection Patrimonie Géologique, co-ed.). (Publications Scientifiques du Muséum/Biotope).

GARASSINO, A. & B. GIRONI, 2006. *Coleia boboi* n. sp. (Crustacea, Decapoda, Eryonoidea) from the Late Triassic (Rhaetian) of Monte Verzegnis (Udine, NE Italy). Atti della Società italiana di Scienze naturali e del Museo civico di Storia naturale in Milano, **147** (1): 93-102.

GARASSINO, A. & D. HARPER, 2004. The macruran decapod crustaceans from the Stolle Klint Clay (Upper Paleocene) of Fur Island (Denmark). Atti della Società italiana di Scienze naturali e del Museo civico di Storia naturale in Milano, **145** (2): 329-336.

GARASSINO, A. & S. L. JAKOBSEN, 2005. *Morscrangon acutus* n. gen. n. sp. (Crustacea: Decapoda: Caridea) from the Fur Formation (early Eocene) of the Islands of Mors and Fur (Denmark). Atti della Società italiana di Scienze naturali e del Museo civico di Storia naturale in Milano, **146** (1): 95-107.

GARASSINO, A. & M. KROBICKI, 2002. *Galicia marianae* n. gen., n. sp. (Crustacea, Decapoda, Astacidea) from the Oxfordian (Upper Jurassic) of the Southern Polish Uplands. Bulletin of the Mizunami Fossil Museum, **29**: 51-59.

GARASSINO, A. & G. PASINI, 2002. Studies on Permo-Trias of Madagascar. 5. *Ambilobeia karojoi* n. gen., n. sp. (Crustacea, Decapoda) from the Lower Triassic (Olenekian) of Ambilobè region (NW Madagascar). Atti della Società italiana di Scienze naturali e del Museo civico di Storia naturale in Milano, **143** (1): 95-104.

GARASSINO, A., G. PASINI & D. B. DUTHEIL, 2006. *Cretapenaeus berberus* n. gen., n. sp. (Crustacea: Decapoda: Penaeidae) from the Late Cretaceous (Cenomanian) of southeastern Morocco. Atti della Società italiana di Scienze naturali e del Museo civico di Storia naturale in Milano, **147** (1): 3-17.

GARASSINO, A. & R. RIGO, 2008. *Pseudoglyphea friulana* n. sp. (Decapoda, Astacidea, Mecochiridae) from the Upper Triassic (Carnian) of Dogna (Udine, Friuli-Venezia Giulia, NE Italy). Atti della Società italiana di Scienze naturali e del Museo civico di Storia naturale in Milano, **149**: 69-76.

GARASSINO, A. & G. SCHWEIGERT, 2004. *Cycleryon wulfi* n. sp. (Crustacea: Decapoda: Eryonidae) from the Upper Jurassic of Eichstätt (Bavaria, S Germany). Atti della Società italiana di Scienze naturali e del Museo civico di Storia naturale in Milano, **145** (2): 243-250.

— — & — —, 2006. *Cretasergestes sahelalmae* n. gen., n. sp. (Crustacea, Decapoda, Sergestoidea) and *Cancrinos libanensis* n. sp. (Crustacea, Decapoda, Palinuroidea) from the Upper Cretaceous (Cenomanian) of Lebanon. Atti della Società italiana di Scienze naturali e del Museo civico di Storia naturale in Milano, **147** (1): 69-78, 6 figs.

GARASSINO, A. & G. TERUZZI, 1990. The genus *Aeger* Münster, 1839 in the Sinemurian of Osteno in Lombardy (Crustacea, Decapoda). Atti della Società italiana di Scienze naturali e del Museo civico di Storia naturale in Milano, **131** (5): 105-136.

— — & — —, 1993. A new decapod crustacean assemblage from the Upper Triassic of Lombardy (N. Italy). Paleontologia Lombarda, (nuova serie) **1**: 1-27, 5 pls.

— — & — —, 1995a. I crostacei decapodi del Cretacico Inferiore di Vernasso (Udine, NE Italia). Gortania – Atti del Museo Friulano di Storia Naturale, **16**: 77-88.

— — & — —, 1995b. Studies on Permo-Trias of Madagascar. 3. The decapod crustaceans of the Ambilobè region (NW Madagascar). Atti della Societa italiana di Scienze naturali e del Museo civico di Storia naturale in Milano, **134** (1): 85-113.

— — & — —, 1996. The genera *Longitergite* nov. and *Bannikovia* nov. in the Lower Miocene of N Caucasus (Russia) (Crustacea, Decapoda). Atti della Società italiana di Scienze naturali e del Museo civico di Storia naturale in Milano, **136** (1): 3-14.

GARASSINO, A., G. TERUZZI & F. DALLA VECCHIA, 1996. The macruran decapod crustaceans of the Dolomia di Forni (Norian, Upper Triassic) of Carnia (Udine, NE Italy). Atti della Società italiana di Scienze naturale e del Museo civico di Storia naturale in Milano, **136**: 15-60.

GARASSINO, A., S. YANBIN, F. R. SCHRAM & R. S. TAYLOR, 2002. *Yongjiacaris zhejiangensis* n. gen. n. sp. (Crustacea, Decapoda) from the Barremian (Lower Cretaceous) of Zhejiang Province (China). Bulletin of the Mizunami Fossil Museum, **29**: 73-80.

GARTH, J. S., 1958. Brachyura of the Pacific coast of America. Oxyrhyncha. Allan Hancock Pacific Expeditions, **21** (1-2): 1-499, pls. A-Z4, 1-55.

GARTH, J. S. & P. K. L. NG, 1985. Notes on the genus *Demania* Laurie, 1906 (Crustacea, Decapoda, Brachyura, Xanthidae). Indo-Malayan Zoology, **2**: 293-308.

GARY, L., 1907. Note sur le terrain éocène des environs de Lagrasse. Bulletin de la Societé d'Études Scientifiques de l'Aude, **18**: 145-172.

GEINITZ, H. B., 1849-1850. Das Quadersandsteingebirge oder Kreidegebirge in Deutschland: 1-99. (Verlag Craz und Gerlach, Freiberg).

— —, 1875. Das Elbthalgebirge in Sachsen und seine fossile Fauna. Palaeontographica, **20**: 1871-1875.

GEMMELLARO, G. G., 1869. Studi paleontologici sulla fauna del calcare à *Terebratula janitor* del Nord di Sicilia, **1**: 11-18, pls. 2, 3. (Palermo).

GEORGIADES-DIKEOULIA, E. & P. MÜLLER, 1984. Palaeoecology of a Messinian dolomite from Iraklion (Greece) with the description of a new crab species. Annales Géologiques des Pays Helléniques, **32**: 237-244, pl. 1.

GERMAR, E. F., 1827. Über die Versteinerungen von Solenhofen. Keferstein's Teutschland Geognostisch-Geologisch dargestellt, **4** (2): 89-110, pl. 1.

GILL, T., 1894. A new Bassalian type of crabs. American Naturalist, **28**: 1043-1045.

— —, 1898. The Crustacean genus *Scyllarides*. Science, New York, **7** (1): 98-99.

GISTL, J. N. F. X. VON *see* VON GISTL, J. N. F. X.

GLAESSNER, M. F., 1924. Über eine neue Miozäne Krabbe und die Dekapodenfauna des Wiener Beckens. Verhandlungen der Geologischen Bundesanstalt, **6**: 109-118.

— —, 1928. Die Dekapodenfauna des österreichischen Jungtertiärs. Jahrbuch der Geologischen Bundesanstalt, **78**: 161-219, figs. 3, 4.

— —, 1929a. Dekapodenstudien. Neues Jahrbuch für Mineralogie, Geologie, und Paläontologie, (Beilage) **63** (B): 137-176, pls. 6-10.

— —, 1929b. Crustacea Decapoda. In: F. J. POMPECKJ (ed.), Fossilium catalogus, **1**: Animalium, (41): 1-464. (W. Junk, Berlin).

— —, 1930. Neues Krebsreste aus der Kreide. Jahrbuch der Preussischen Geologischen Landesanstalt zu Berlin, **51**: 1-7, pl. 1.

— —, 1931. Eine Crustaceenfauna aus den Lunzer Schichten Niederösterreichs. Jahrbuch der k. k. Geologischen Bundesanstalt Wien, **81**: 467-486, pls. 15, 16.

— —, 1932a. Neue Krebsreste aus der Kreide. Jahrbuch der Preussischen Geologischen Landesanstalt, **53**: 577-586, pl. 28.

— —, 1932b. Zwei ungenügend bekannte mesozoische Dekapodenkrebse. *Pemphix sueuri* (Desm.) und *Palaeophoberus suevicus* (Quenstedt). Paläontologische Zeitschrift, **14**: 108-121.

— —, 1933a. Die Krabben der Juraformation. Zentralblatt für Mineralogie, Geologie und Paläontologie, (B) **3**: 178-191.

— —, 1933b. New Tertiary crabs in the collection of the British Museum. Annals and Magazine of Natural History, (10) **12**: 1-28, pls. 1-6.

— —, 1946. Cretaceous Crustacea from Mount Lebanon, Syria. Annals and Magazine of Natural History, London, (11) **12**: 694-707.

— —, 1948. Decapod Crustacea (Callianassidae) from the Eocene of Victoria. Proceedings of the Royal Society of Victoria, (new series) **59**: 1-7.

— —, 1956. Crustacea from the Cretaceous and Eocene of Western Australia. Journal of the Royal Society of Western Australia, **40**: 33-35, 1 pl.

— —, 1960. The fossil decapod Crustacea of New Zealand and the evolution of the order Decapoda. New Zealand Geological Survey Paleontological Bulletin, **31**: 3-63, pls. 1-7.

— —, 1965. Vorkommen fossiler Dekapoden (Crustacea) in Fisch-Schiefern. Senckenbergiana Lethaia, **46** (a): 111-122.

— —, 1969. Decapoda. In: R. C. MOORE (ed.), Treatise on Invertebrate Paleontology, **R** (4) (2): R400-R533, R626-R628. (Geological Society of America, Boulder, Colorado, and University of Kansas Press, Lawrence, Kansas).

— —, 1980. New Cretaceous and Tertiary crabs (Crustacea: Brachyura) from Australia and New Zealand. Transactions of the Royal Society of South Australia, **104**: 171-192.

GLAESSNER, M. F. & V. R. RAO, 1960. A new species of crab from the early Tertiary Fuller's Earth deposits of Kapurdi, Rajasthan, Western India. Records of the Geological Survey of India, **86**: 675-682.

GLAESSNER, M. F. & S. SECRETAN, 1987. Crabes (Crustacea Brachyura) de l'Éocène du Sulaiman Range (Pakistan). Annales de Paléontologie, (Vert.-Invert.) **73**: 273-288.

GOEKE, G. D., 1981. Symethinae, new subfamily, and *Symethis garthi*, new species, and the transfer of *Raninoides ecuadorensis* to *Notosceles* (Raninidae: Brachyura: Gymnopleura). Proceedings of the Biological Society of Washington, **93**: 971-981.

— —, 1985. Decapod Crustacea: Raninidae. Res. Camp. MUSORSTOM I-II. Philippines (1976, 1980), **2**. Mémoires du Muséum National d'Histoire Naturelle, Paris, (A, Zoologie) **133**: 205-228.

GORODISKI, A. & J.-M. REMY, 1959. Sur les Décapodes éocènes du Sénégal occidental. Bulletin de la Société Géologique de France, (7) **1** (3): 315-319, pl. 19a.

GOULD, C., 1857. Description of a fossil crustacean (*Tropifer laevis*) from the Lias Bone-Bed. Quarterly Journal of the Geological Society, **13**: 360-363.

— —, 1859. Description of a new fossil crustacean from the lower Greensand. Quarterly Journal of the Geological Society, **15**: 237-238.

GRAVE, S. DE *see* DE GRAVE, S.

GRAY, J. E., 1845. Description of some new Australian animals. In: E. J. EYRE, Journals of expeditions of discovery into central Australia, etc., **1**: 405-411.

GREGORIO, A. DE *see* DE GREGORIO, A.

GRIFFIN, D. J. G., 1963. *Notomithrax* gen. nov. and the status of the genus *Paramithrax* H. Milne Edwards (Crustacea, Brachyura, Majidae). Transactions of the Royal Society of New Zealand, (Zoology) **3**: 229-237.

GRIPP, K., 1927. Beiträge zur Geologie von Spitzbergen. Abhandlungen aus dem Gebiete der Naturwissenschaften, herausgegeben vom Naturwissenschaftlichen Verein Hamburg, **21** (3-4): 1-38, pls. 1-7.

GROENEVELD, J. C., R. GOÑI & D. LATROUITE, 2006. *Palinurus* species. In: B. F. PHILLIPS (ed.), Lobsters: biology, management, aquaculture and fisheries: 385-411. (Blackwell Publishing, Oxford, U.K.).

GUÉRIN-MÉNEVILLE, F. E., 1832. Notice sur quelques modifications à introduire dans les Notopodes de M. Latreille et établissement d'un nouveau genre dans cette tribu. Annales des Sciences Naturelles, **25**: 283-289.

GUINOT, D., 1967. Recherches préliminaires sur les groupements naturels chez les Crustacés Décapodes Brachyoures. II. Les anciens genres *Micropanope* Stimpson et *Medaeus* Dana. Bulletin du Muséum National d'Histoire Naturelle, Paris, (2) **39** (2): 345-374.

— —, 1969. Sur divers Xanthidae notamment sur *Actaea* De Haan et *Paractaea* gen. nov. (Crustacea Decapoda Brachyura). Cahiers du Pacifique, **13**: 223-267, figs. 1-36.

— —, 1976. Constitution de quelques groupes naturels chez les Crustacés Décapodes Brachyoures. I. La superfamille des Bellioidea Dana et trois sous-familles de Xanthidae (Polydectinae Dana, Trichiinae De Haan, Actaeinae Alcock). Mémoires du Muséum National d'Histoire Naturelle, Paris, (A) **97**: 1-308, pls. 1-19.

— —, 1977. Propositions pour une nouvelle classification des Crustacés Décapodes Brachyoures. Comptes Rendus Hebdomadaires des Séances de l'Académie des Sciences, Paris, (D) **285**: 1049-1052.

— —, 1986. Description d'un Crabe cavernicole aveugle de Nouvelle-Bretagne (Papouasie Nouvelle-Guinée), *Trogloplax joliverti* gen. nov. sp. nov., et établissement d'une sous-famille nouvelle, Trogloplacinae subfam. nov. Comptes Rendus Hebdomadaires des Séances de l'Academie des Sciences, Paris, (3) **303**: 307-312.

— —, 1990. Établissement du genre *Garthiope* gen. nov., ses relations avec le genre *Coralliope* Guinot, 1967, et leurs affinités avec les Trapeziidae sensu lato (Crustacea Decapoda Brachyura). Bulletin du Muséum National d'Histoire Naturelle, (A, Zoologie) (Biologie et écologie animaux) **12** (2): 469-487.

— —, 1991. Établissement de la famille des Poupiniidae pour *Poupina hirsuta* gen. nov., sp. nov. de Polynésie (Crustacea Decapoda Brachyura Homoloidea). Bulletin du Muséum National d'Histoire Naturelle, Paris, (4) **12** [1990]: 577-605.

— —, 1993. Données nouvelles sur les Raninoidea De Haan, 1841 (Crustacea Decapoda Brachyura Podotremata). Comptes Rendus Académie des Sciences, Paris, (Sciences de la Vie) **316**: 1324-1331.

GUINOT, D., A. DE ANGELI & A. GARASSINO, 2007. *Hebertides jurassica* n. gen., n. sp. (Crustacea, Decapoda, Brachyura) from the Middle Jurassic (Bathonian) of Normandy (France). Atti della Società italiana di Scienze naturali e del Museo civico di Storia naturale in Milano, **148** (2): 241-260.

— —, — — & — —, 2008. Marocarcinidae, a new eubrachyuran family, and *Marocarcinus pasinii* n. gen., n. sp. from the Upper Cretaceous (Cenomanian-Turonian) of Gara Sbaa, southeastern Morocco (Crustacea, Decapoda, Brachyura). Atti della Società italiana di Scienze naturali e del Museo civico di Storia naturale in Milano, **148** (2): 25-36.

GUINOT, D. & E. MACPHERSON, 1987. Révision du genre *Pilumnoides* Lucas, 1844, avec description de quatre espèces nouvelles et création de Pilumnoidinae subfam. nov. (Crustacea Decapoda Brachyura). Bulletin du Muséum National d'Histoire Naturelle, Paris, (4) **9** (A) (1): 211-247.

GUINOT, D. & B. RICHER DE FORGES, 1995. Crustacea Decapoda Brachyura: révision de la famille des Homolidae De Haan, 1839. In: A. CROSNIER (ed.), Résultats des Campagnes MUSORSTOM, **13**. Mémoires du Muséum National d'Histoire Naturelle, Paris, **163**: 283-517.

GUINOT, D. & M. TAVARES, 2001. Une nouvelle famille de crabes du Crétacé et la notion de Podotremata Guinot, 1977 (Crustacea, Decapoda, Brachyura). Zoosystema, **23**: 507-546.

— — & — —, 2003. A new subfamilial arrangement for the Dromiidae De Haan, 1833, with diagnoses and descriptions of new genera and species (Crustacea, Decapoda, Brachyura). Zoosystema, **25**: 43-129.

GUINOT, D., F. J. VEGA & B. VAN BAKEL, 2008. Cenomanocarcinidae n. fam., a new Cretaceous podotreme family (Crustacea, Decapoda, Brachyura, Raninoidea), with comments on related families. Geodiversitas, **30** (4): 681-719.

GUPPY, R. J. L., 1909. Preliminary notice of a discovery of fossils in the Tamana District, Trinidad. Bulletin of the Agricultural Department, Trinidad, **1909**: 5-6, 1 pl.

HAAN, W. DE *see* DE HAAN, W.

HAAS, H., 1889. Ueber *Podocrates* und *Homarus* aus dem Mitteloligozän von Itzehoe. Mitteilungen aus dem Mineralogischen Institut der Universität Kiel, **1**: 88-97.

HAIME, J., 1855. Notice sur la géologie d l'ile de Majorque. Bulletin de la Societé Géologique de France, (2) **2**: 734-750, pl. 15.

HALL, J. & F. B. MEEK, 1855. Descriptions of new species of fossils from the Cretaceous Formations of Nebraska. Memoirs of the American Academy of Arts and Sciences, (new series) **5**: 379-411, pls. 1-8.

HARBORT, E., 1905. Die Fauna der Schaumburg-Lippeschen Kreidemulde. Abhandlungen der Preussischen Geologischen Landesanstalt, (N. F.) **14**: 10-22.

HART, C. W. & R. B. MANNING, 1986. Two new shrimps (Procarididae and Agostocaridae, new family) from marine caves of the western North Atlantic. Journal of Crustacean Biology, **6** (3): 408-416.

HARVEY, A. W., 1998. New names for Miocene hermit crabs (Decapoda, Anomura, Paguridae). Crustaceana, **71**: 119-121.

HASWELL, W. A., 1882. Catalogue of the Australian stalk- and sessile-eyed Crustacea: 1-324 + addenda et corrigenda, pls. 1-4. (The Australian Museum; F. W. White, Sydney).

HATAI, K. & T. KOTAKA, 1970. A problematica from the Nishikurosawa Formation, Oga Peninsula, Akita Prefecture, northeast Honshu, Japan. Saito Ho-on Kai Museum Research Bulletin, **39**: 1-5.

HAWORTH, A. H., 1825. A new binary arrangement of the macrurous Crustacea. Philosophical Magazine and Journal, **65**: 183-184.

HÉE, A., 1924. Catalogue critique des Crustacés jurassiques du Calvados et de l'Orne. Bulletin de la Societé Linnéenne de Normandie, (7) **6**: 126-157, pls. 3-6.

HELLER, C., 1861. Beiträge zur Crustaceen-Fauna des Rothen Meeres. Sitzungsberichte der Bayerischen Akademie der Wissenschaften zu München, (Mathematisch-physikalische Klasse) **43**: 297-374.

— —, 1862. Neue Crustaceen, gesammelt während der Weltumseglung der k. k. Fregatte Novara: zweiter vorläufiger Bericht. Verhandlungen der Kaiserlich-Königlichen Zoologisch-Botanischen Gesellschaft in Wien, **12**: 519-528.

— —, 1865. Reise der oesterreichischen Fregatte "Novara" um die Erde in den Jahren 1857-1859 unter den Befehlen des Commodore, B. von Wullerstorf-Urbair. Die Crustaceen, **1**. (Vienna).

HENDERSON, J. R., 1886. The decapod and schizopod Crustacea of the Firth of Clyde. Proceedings and Transactions of the Natural History Society of Glasgow, **3**: 315-353.

— —, 1888. Report on the Anomura collected by H. M. S. Challenger during the years 1873-1876. Report on the Scientific Results of the Voyage of HMS Challenger, (Zoology) **27**: i-xi, 1-221, pls. 1-21.

— —, 1893. A contribution to Indian carcinology. Transactions of the Linnean Society of London, (2, Zoology) **5** (1): 325-458, pls. 36-40.

HERBST, J. F. W., 1782-1804. Versuch einer Naturgeschichte der Krabben und Krebse nebst einer systematischen Beschreibung ihrer verschiedenen Arten, **1** [1782-1790]: 1-274, pls. 1-21; **2** [1791-1796]: i-viii, iii, iv, 1-225, pls. 22-46; **3** [1799-1804]: 1-66, pls. 47-50. (G. A. Lange, Berlin; J. C. Fuessly, Zürich).

HOBBS, H. H., JR., 1942. The crayfishes of Florida. University of Florida Publications, (Biological Science Series) **3** (2): 1-179, figs. 1-24.

HOEPEN, E. C. N. VAN *see* VAN HOEPEN, E. C. N.

HOLLAND, F. D., JR. & A. CVANCARA, 1958. Crabs from the Cannonball Formation (Paleocene) of North Dakota. Journal of Paleontology, **32**: 495-505.

HOLMES, S. J., 1904. Synopsis of California stalk-eyed Crustacea. Occasional Papers of the California Academy of Science, **7**: 1-262.

HOLTHUIS, L. B., 1955. The recent genera of the caridean and stenopodidean shrimps (class Crustacea, order Decapoda, supersection Natantia) with keys for their determination. Zoologische Verhandelingen, Leiden, **26**: 1-157.

— —, 1960. Preliminary descriptions of one new genus, twelve new species and three new subspecies of scyllarid lobster (Crustacea Decapoda Macrura). Proceedings of the Biological Society of Washington, **73**: 147-154.

— —, 1961. Notes on American Albuneidae (Crustacea Decapoda, Anomura) with the description of a new genus and species. Proceedings Koninklijke Nederlandse Akademie van Wetenschappen, (C, Biological and Medical Sciences) **64**: 21-36.

— —, 1987. *Necora*, a new genus of European swimming crabs (Crustacea Decapoda, Portunidae) and its type species, *Cancer puber* L., 1767. Zoologische Medelingen, Leiden, **61** (1): 1-14.

— —, 1991. Marine lobsters of the world: an annotated and illustrated catalogue of species of interest to fisheries known to date. FAO Fisheries Synopsis, 125; **13**: i-viii, 1-292. (Food and Agriculture Organization of the United Nations, Rome).

— —, 2002. *Galearctus* the Indo-Pacific scyllarine lobsters (Crustacea, Decapoda, Scyllaridae). Zoosystema, **14** (3): 499-683.

HOLTHUIS, L. B. & R. B. MANNING, 1990. Crabs of the subfamily Dorippinae MacLeay, 1838, from the Indo-West Pacific region (Crustacea: Decapoda: Dorippidae). Researches on Crustacea, (special number) **3**: 1-157. (The Carcinolocical Society of Japan, Tokyo).
HOUŠA, V., 1957 [imprint 1956]. *Bechleja inopinata* n. g., n. sp., ein neuer Krebs aus dem böhmischen Tertiär (Decapoda, Palaemonidae). Sborník Ústředního Ústavu Geologické, (Oddíl Paleontologický) **23**: 365-377.
— —, 1963. Parasites of Tithonian decapod crustaceans (Štramberk, Moravia). Sborník Ústředního Ústavu Geologické, Paleontologie, **28**: 101-114.
HU, C.-H., 1981. Studies on Cenozoic fossil crabs from Taiwan Island. Proceedings of the Geological Society of China, **24**: 56-74, pls. 1-4.
— —, 1983. Some Cenozoic fossil crabs from Taiwan. Proceedings of the Geological Society of China, **26**: 76-98, pls. 1-4.
— —, 1984a. Discovery fossil lobster from the Kueichulin Formation (Miocene), southern Taiwan. Taiwan Museum Annals, **26**: 128-136.
— —, 1984b. Cenozoic crab fossils from Taiwan Island. Petroleum Geology of Taiwan, **20**: 181-198, pls. 1-7.
HU, C.-H. & H.-J. TAO, 1979. Studies on Quaternary fossil crabs and associated faunas from Hainan Island, southern China. Proceedings of the Geological Society of China, **22**: 145-158.
— — & — —, 1985. Tertiary crustaceans fossil from Taiwan. Petroleum Geology of Taiwan, **21**: 239-259, pls. 1-7.
— — & — —, 1996. Crustacean fossils of Taiwan: 1-228. (Taipei, Taiwan, Republic of China).
— — & — —, 1999. Cenozoic crab fossils from Taiwan island. Petroleum Geology of Taiwan, **33**: 217-240.
— — & — —, 2000. Crustacean fossils from southern Taiwan. Petroleum Geology of Taiwan, **34**: 105-195.
— — & — —, 2004. Studies on the Neogene crabs from south-western foot hills of Taiwan. Acta Palaeontologica Sinica, **43** (4): 537-555.
HUXLEY, T. H., 1879. On the classification and the distribution of the crayfishes. Proceedings of the Scientific Meetings of the Zoological Society of London, **1878**: 752-788.
HYDEN, F. M. & J. FOREST, 1980. An *in situ* hermit crab from the early Miocene of southern New Zealand. Palaeontology, **23** (2): 471-474.
IHLE, J. E. W., 1913. Die Decapoda Brachyura der Siboga-Expedition. I. Dromiacea. Siboga Expeditie, **39** (b): 1-96, pls. 1-4.
ILYIN, I. V., 2005. Cretaceous and Paleogene decapod crustaceans of the western part of northern Eurasia: 1-296. (State University Press, Moscow). [In Russian.]
ILYIN, I. V. & A. S. ALEKSEEV, 1998. New crab species (Decapoda, Brachyura) from the Lower Cretaceous of south-west Crimea. Paleontological Journal, **32**: 592-595. [Translated from Paleontologicheskii Zhurnal.]
IMAIZUMI, R., 1939. *Palaeocarpilius laevis* sp. nov. from the Tappocho Limestone of Saipan Mariana Group. Jubilee Publication in the Commemoration of Professor, H. Yabe, M. I. A. Sixtieth Birthday, **1**: 221-226, pl. 14.
— —, 1951. *Trachycarcinus huziokai* n. sp. from Yamagata Prefecture. Short Papers, IGPS [Short Papers from the Institute of Geology and Palaeontology, Tohoku University], **3**: 33-40, pl. 6.
— —, 1952a. *Callianassa tayamai* n. sp. from Saipan. Short Papers, IGPS [Short Papers from the Institute of Geology and Palaeontology, Tohoku University], **5**: 77-83, pls. 10, 11.

— —, 1952b. A Miocene crab, *Tymolus kamadai* n. sp. from the Numanouchi Formation of the Joban coal-field. Transactions and Proceedings of the Palaeontological Society of Japan, (new series) **7**: 201-204.
— —, 1952c. *Mursia takahashii* n. sp. (Brachyura) from Goishi in the Sendai Area. Short Papers, IGPS [Short Papers from the Institute of Geology and Palaeontology, Tohoku University], **4**: 88-95.
— —, 1952d. A Miocene crab, *Hyas tsuchidai* n. sp. from the Wakkanai Formation of Teshio Province, Hokkaido. Transactions and Proceedings of the Palaeontological Society of Japan, (new series) **5**: 179-183.
— —, 1957a. Three new Miocene species of *Callianassa* from Nagano Prefecture. Transactions and Proceedings of the Palaeontological Society of Japan, (new series) **27**: 81-85, pl. 14.
— —, 1957b. A Miocene fossil crab, *Paratymolus yabei* n. sp. from Nagano Prefecture. Transactions and Proceedings of the Palaeontological Society of Japan, (new series) **25**: 26-30, pl. 5.
— —, 1958. An interesting crustacean remain, *Ctenocheles sujakui* n. sp. from the Paleogene of Kyushu Japan. Transactions and Proceedings of the Palaeontological Society of Japan, (new series) **32**: 299-304, pl. 44.
— —, 1959a. *Callianassa bona* n. sp. from near Sendai, Miyagi Prefecture. Japanese Journal of Geology and Geography, **30**: 31-37, pl. 4.
— —, 1959b. A fossil crab, *Hexapus nakajimai* n. sp. from Joban coal field. Japanese Journal of Geology and Geography, **30**: 25-30, pl. 2.
— —, 1961. A critical review and systematic descriptions of known and new species of *Carcinoplax* from Japan. Science Reports of the Tohoku University, (2, Geology) **32** (2): 155-193, pls. 12-22.
— —, 1962. Miocene *Cancer* (Brachyura) of Japan. Science Reports of the Tohoku University, (2, Geology) (special volume) **5**: 223-246, pl. 40.
— —, 1965. Miocene *Macrocheira* from Japan. Researches on Crustacea, **2**: 27-36, pl. 4.
JACQUINOT, C. H. & H. LUCAS, 1853. Voyage au Pole Sud et dans l'Océanie sur les corvettes Astrolabe et La Zelée, (Zoologie) **3**, Crustacés: 1-107. (Gide et Baudry, Paris).
JAGT, J. W. M., B. W. M. VAN BAKEL & R. H. B. FRAAIJE, 2007. *Palaeodromites crypticus*, a new Early-Middle Cenomanian dynomenid crab (Crustacea, Decapoda) from southern Belgium. Bulletin de l'Institut Royal des Sciences Naturelles de Belgique, (Sciences de la Terre) **77**: 77-82, 2 figs.
JAGT, J. W. M., J. S. H. COLLINS & R. H. B. FRAAYE, 1991. A new late Maastrichtian xanthid crab from southern Limburg (the Netherlands). Cretaceous Research, **12**: 553-560.
— —, — — & — —, 1993. A new early Palaeocene genus of raninid crab (Crustacea, Decapoda) from Denmark, southern Sweden and the Netherlands. Contributions to Tertiary and Quaternary Geology, **30** (3-4): 177-182, 3 figs.
JAKOBSEN, S. L. & J. S. H. COLLINS, 1979. Decapod Crustacea from the Palaeocene of Zealand, Denmark. Proceedings of the Geologists Association, **90** (2): 61-64.
— — & — —, 1997. New Middle Danian species of anomuran and brachyuran crabs from Fakse, Denmark. Bulletin of the Geological Society of Denmark, **44**: 89-100.
JANSSEN, A. W. & P. MÜLLER, 1984. Miocene Decapoda and Mollusca from Ramsel (province of Antwerpen, Belgium), with a new crab genus and a new cephalopod species. Scripta Geologica, **75**: 1-26, pls. 1-5.
JENKINS, R. J. F., 1972. *Metanephrops*, a new genus of late Pliocene to Recent lobsters (Decapoda, Nephropidae). Crustaceana, **22**: 161-177.
— —, 1974. A new spider-crab from the Miocene of New Zealand. Palaeontology, **17**: 869-877.

——, 1977. A new fossil homolid crab (Decapoda, Brachyura), Middle Tertiary, southeastern Australia. Transactions of the Royal Society of South Australia, **101**: 1-10.

——, 1985. Fossil spider crabs from Australia. South Australia Department Mines and Energy, (special publication) **5**: 145-165.

JIMBÔ, K., 1894. Beiträge zur Kenntnis der Fauna der Kreideformation von Hokkaidô. Paläontologische Abhundlungen, (N. F.) **2** (3): 149-194, pls. 17-25.

JOHNSON, J. Y., 1861. Description of a new species of *Cancer* obtained at Madeira. Proceedings of the Zoological Society of London, **1861**: 240-242.

JOLEAUD, L. & T. HSU, 1935. Crustacés Décapodes du Crétacé de Tanout. Archives du Muséum National d'Histoire Naturelle, Paris, **13** (6): 100-110.

JUX, U., 1971. Ein Brachyuren-Rest aus der Oberkreide Afghanistans. Paläontologische Zeitschrift, **45**: 154-166.

KARASAWA, H., 1989. Decapod crustaceans from the Miocene Mizunami Group, central Japan. Part 1. Superfamily Thalassinoidea, Leucosioidea and Grapsidoidea. Bulletin of the Mizunami Fossil Museum, **16**: 1-28.

——, 1990. Decapod crustaceans from the Miocene Mizunami Group, central Japan, Pt. 2. Oxyrhyncha, Cancridea, and Brachyrhyncha. Bulletin of the Mizunami Fossil Museum, **17**: 1-34, pls. 1-8.

——, 1991. Decapod crustaceans from the Miocene Mizunami Group, central Japan. Part 3. Decapod crustacean assemblage and paleoecology, with description of two species. Bulletin of the Mizunami Fossil Museum, **18**: 1-18.

——, 1992a. Fossil decapod crustaceans from the Manda Group (Middle Eocene), Kyushu, Japan. Transactions and Proceedings of the Palaeontological Society of Japan, (new series) **167**: 1247-1258.

——, 1992b. The crab, *Ozius collinsi* sp. nov. (Crustacea: Decapoda: Xanthoidea) from the Miocene Katsuta Group, southwest Japan. Tertiary Research, **14** (1): 19-24.

——, 1993. Cenozoic decapod Crustacea from southwest Japan. Bulletin of the Mizunami Fossil Museum, **20**: 1-92, 24 pls.

——, 1998a. Two new species of Decapoda (Crustacea) from the Upper Cretaceous Izumi Group, Japan. Paleontological Research, **2**: 217-223.

——, 1998b. *Typilobus kishimotoi*, a new leucosiid crab (Crustacea: Decapoda: Brachyura) from the Miocene Katsuta Group, Japan. Proceedings of the Biological Society of Washington, **111** (1): 97-101.

——, 2000a. A new raninid crab (Crustacea: Decapoda: Brachyura) from the Kase Formation (Lower Miocene) of Kyushu, Japan. Proceedings of the Biological Society of Washington, **113** (3): 806-809.

——, 2000b. *Medorippe tanabei*, a new species of Miocene dorippid crab (Crustacea: Decapoda: Brachyura) from the Katsuta Group, west Honshu, Japan. Proceedings of the Biological Society of Washington, **113** (3): 810-814.

——, 2002. First record of a Miocene matutid crab (Crustacea: Decapoda: Brachyura) from Japan. Bulletin of the Mizunami Fossil Museum, **29**: 93-94.

——, 2008. *Ebalia noetlingi*, a replacement name for *Ebalia tuberculata* Noetling, 1901, non *Ebalia tuberculata* Miers, 1881 (Crustacea: Decapoda: Brachyura: Leucosiidae). Paleontological Research, **12**: 419.

KARASAWA, H. & Y. FUDOUJI, 2000. Palaeogene decapod Crustacea from the Kishima and Okinoshima Groups, Kyushu, Japan. Paleontological Research, **4** (4): 239-253.

KARASAWA, H. & Y. GODA, 1996. Two species of decapod crustaceans from the Middle Pleistocene Atsumi Group, Japan. Science Report of the Toyohashi Museum of Natural History, **6**: 1-4.

KARASAWA, H. & H. HAYAKAWA, 2000. Additions to Cretaceous decapod crustaceans from Hokkaido, Japan – Part 1. Nephropidae, Micheleidae and Galatheidae. Paleontological Research, **4** (2): 139-145.

KARASAWA, H. & K. INOUE, 1992. Decapod crustaceans from the Miocene Kukinaga Group, Tanegashima Island, Kyushu, Japan. Tertiary Research, **14** (2): 73-96.

KARASAWA, H. & H. KATO, 1996. *Daldorfia* Rathbun, 1904 (Crustacea, Decapoda) from the Neogene of Japan. Proceedings of the Biological Society of Washington, **109** (1): 44-52.

— — & — —, 2003. The family Goneplacidae MacLeay, 1838 (Crustacea: Decapoda: Brachyura): systematics, phylogeny, and fossil records. Paleontological Research, **7** (2): 129-151.

— — & — —, 2007. New prosopid crabs (Crustacea, Decapoda, Brachyura) from the Upper Jurassic Torinosu Group, Shikoku, Japan. Memoire della Società italiana di Scienze naturali e del Museo civico di Storia naturale in Milano, **35** (2): 62-65.

KARASAWA, H., H. KATO, T. KASE, Y. MAAC-AGUILAR, Y. KURIHARA, H. HAYASHI & K. HAGINO, 2008. Neogene and Quaternary ghost shrimps and crabs (Crustacea: Decapoda) from the Philippines. Bulletin of the National Museum of Natural History, (C) **34**: 51-76.

KARASAWA, H., H. KATO & K. TERABE, 2006. A new member of the family Prosopidae (Crustacea: Decapoda: Brachyura) from the Lower Cretaceous of Japan. Revista Mexicana de Ciencias Geológicas, **23** (3): 344-349.

KARASAWA, H. & S. KISHIMOTO, 1996. Two new species of decapod crustaceans from the Katsuta Group (middle Miocene), Japan. Bulletin of the Mizunami Fossil Museum, **23**: 35-37.

KARASAWA, H. & K. KITAMURA, 2003. A first record of *Pugettia* Dana (Crustacea: Decapoda: Brachyura) from the Pliocene of Japan. Bulletin of the Mizunami Fossil Museum, **30**: 61-62.

KARASAWA, H. & T. NOBUHARA, 2008. Decapoda and Isopoda (Crustacea) from the Pliocene Shimajiri Group in the Miyako-jima Island, Ryukyus, Japan. Bulletin of the Mizunami Fossil Museum, **34**: 23-30.

KARASAWA, H. & M. OHARA, 2009. *Ponotus shirahamensis*, a new genus and species of raninid crabs (Decapoda, Brachyura) from the Miocene Tanabe Group of Japan. Boletín de la Sociedad Geológica Mexicana, **61**: 199-202.

KARASAWA, H., M. OHARA & H. KATO, 2008. New records for Crustacea from the Arida Formation (Lower Cretaceous, Barremian) of Japan. Boletín de la Sociedad Geológica Mexicana, **60**: 101-110.

KARASAWA, H., T. SAKUMOTO & H. TAKAYASU, 1992. A new portunid genus, *Itoigawaia*, from the Miocene of southwest Japan. Bulletin of the Mizunami Fossil Museum, **19**: 455-477.

KARASAWA, H. & C. E. SCHWEITZER, 2004. Revision of the genus *Glyphithreus* Reuss, 1859 (Crustacea: Decapoda: Brachyura: Xanthoidea) and recognition of a new genus. Paleontological Research, **8** (3): 143-154.

— — & — —, 2006. A new classification of the Xanthoidea *sensu lato* (Crustacea: Decapoda: Brachyura) based on phylogenetic analysis and traditional systematics and evaluation of all fossil Xanthoidea *sensu lato*. Contributions to Zoology, **75** (1/2): 23-73.

KARASAWA, H., C. E. SCHWEITZER & R. M. FELDMANN, 2008. Revision of the Portunoidea Rafinesque, 1815 (Decapoda: Brachyura) with emphasis on the fossil genera and families. Journal of Crustacean Biology, **28**: 82-127.

KARASAWA, H., F. TAKAHASHI, E. DOI & H. ISHIDA, 2003. First notice of the family Coleiidae Van Straelen (Crustacea: Decapoda: Eryonoidea) from the upper Triassic of Japan. Paleontological Research, **7** (4): 357-362.

KARASAWA, H. & T. TANAKA, 2005. A first notice of *Acmaeopleura* Stimpson (Crustacea: Decapoda: Brachyura) from the Miocene of Japan. Bulletin of the Mizunami Fossil Museum, **32**: 95-96.

KATO, H., 1996. Miocene decapod Crustacea from the Chichibu Basin, central Japan. Transactions and Proceedings of the Palaeontological Society of Japan, (new series) **183**: 500-521.

— —, 2002. Fossil crabs (Crustacea: Decapoda: Brachyura) from the latest Miocene Senhata Formation, Boso Peninsula, Japan. Paleontological Research, **6** (2): 211-217.

— —, 2005. A new pinnotherid crab (Decapoda: Brachyura: Pinnotheridae) from the Miocene Niijukutoge Formation, northeast Japan. Paleontological Research, **9** (1): 73-78.

KATO, H. & Y. HIKIDA, 2002. A new fossil *Cancer* (Decapoda: Brachyura: Cancridae) from the Middle Miocene of northern Hokkaido, Japan. Bulletin of Nakagawa Museum of Natural History, **5**: 47-52.

KATO, H. & H. KARASAWA, 1994. *Minohellenus macrocheilus* sp. nov. (Decapoda: Crustacea) from the Oligocene Ashiya Group, Kyushu, Japan. Bulletin of the Kitakyushu Museum of Natural History, **13**: 51-58.

— — & — —, 2006. New nephropid and glypheid lobsters from the Mesozoic of Japan. Revista Mexicana de Ciencias Geológicas, **23** (3): 338-343.

KATO, H. & A. KOIZUMI, 2001. A new fossil geryonid crab (Crustacea: Decapoda: Brachyura) from the Late Pliocene to Early Pleistocene of Japan. Proceeding of the Biological Society of Washington, **114**: 172-177.

KENSLEY, B., 1985. The faunal deposits of a late Pleistocene raised beach at Milnerton, Cape Province, South Africa. Annals of the South African Museum, **95**: 111-122.

KENSLEY, B. & A. B. WILLIAMS, 1990. *Axiopsis eximia*, a new thalassinidean shrimp (Crustacea, Decapoda, Axiidae) from the Middle Eocene of South Carolina. Journal of Paleontology, **64**: 798-802.

KESLING, R. V., 1958. Fossil crabs from Guam. Contributions from the Museum of Paleontology, University of Michigan, **14**: 207-263.

KISHINOUYE, K., 1926. Two rare and remarkable forms of macrurous Crustacea from Japan. Annotationes Zoologicae Japonenses, **11**: 63-70.

KITCHIN, F. L., 1908. The invertebrate fauna and palaeontological relations of the Uitenhage Series. Annals of the South African Museum, **7**: 212-268.

KNEBEL, R. VON *see* VON KNEBEL, R.

KÖNIG, C., 1825. Icones fossilium sectiles: 1-4, 19 pls. (G. B. Sowerby, London).

KÖNIG, H., 1920. Zur Kenntnis des untern Trochitenkalkes im nördlichen Kraichgau. Sitzungsberichte der Heidelberger Akademie der Wissenschaften, Heidelberg, **13**: 9-48.

KOOSER, M. A. & W. N. ORR, 1973. Two new decapod species from Oregon. Journal of Paleontology, **47**: 1044-1046.

KORNFIELD, I., A. B. WILLIAMS & R. S. STENECK, 1995. Assignment of *Homarus capensis* (Herbst, 1792), the Cape lobster of South Africa, to the new genus *Homarinus* (Decapoda: Nephropidae). U.S. National Marine Fisheries Service Fishery Bulletin, **93** (1): 97-102.

KOSSMANN, R., 1880. Zoologische Ergebnisse einer Reise in die Küstengebiete des Rothen Meeres, **2** (1) (3), Malacostraca. Zoologische Ergebnisse im Aufträge der Königlichen Academie der Wissenschaften zu Berlin, **1880**: 67-140, pls. 4-15.

KOWALSKI, J., 1923. Description d'une nouvelle espèce de Crabe fossile provenant des Nouvelles-Hébrides. Bulletin de la Societé Scientifique Naturelle Ouest, (4) **3**: 13-24.

KRAUSE, P. G., 1891. Die Decapoden des norddeutschen Jura. Zeitschrift der Deutschen Geologischen Gesellschaft, **43**: 171-225.

— —, 1908. Über Diluvium, Tertiär, Kreide und Jura in der Heilsberger Tiefbohrung. Jahrbuch der Königlichen Preussischen Geoligischen Landesanstalt und Bergakademie in Berlin, **29**: 185-326, pls. 3-8.

KUHN, O., 1952. Neue Crustacea Decapoda und Insecta aus dem Untersten Lias & von Nordfranken. Palaeontographica, Stuttgart, **101**: 153-166.

LAHUSEN, J., 1894. Über die russischen Krebsreste aus den jurassischen Ablagerungen und derunteren Wolgastufe. Verhandlungen der Russischen Kaiserlichen Minerologischen Gesellschaft, (2) **31**: 313-324.

LAMARCK, J. B. P. A., 1801. Système des animaux sans vertébrés, ou tableau général des classes, des ordres et des genres de ces animaux; présentant leurs caractères essentiels et leurs distribution, d'après la considération de leurs rapports naturels et de leur organisation, et suivant l'arrangement établi dans les galeries du Muséum d'Histoire Naturelle, parmi leurs dépouilles conservées; précédé du discours d'ouverture du cours de zoologie, donné dans le Muséum national d'Histoire naturelle l'an 8 de la Republique: 1-432. (Déterville, Paris).

— —, 1818. Histoire naturelle des animaux sans vertébrés, présentant les caractères généraux et particuliers de ces animaux, leur distribution, leurs classes, leurs familles, leurs genres, et la citation des principales espèces qui s'y rapportent; précédée d'une introduction offrant la détermination des caractères essentiels de l'animal, sa distinction du végétal et des autres corps naturels, enfin, l'exposition des principes fondamentaux de la zoologie (éd. 5): 1-612.

LANGENHAN, A., 1910. Die Versteinerungen der deutschen Trias. (Liegnitz, 1903; and Friedrichsroda, 1910).

LARGHI, C., 2002. *Mithracia oppionii* sp. nov. (Crustacea, Decapoda, Brachyura) from the Eocene of Chiampo (Vicenza, Italy). Bulletin of the Mizunami Fossil Museum, **29**: 61-68, figs. 1-5.

— —, 2004. Brachyuran decapod Crustacea from the Upper Cretaceous of Lebanon. Journal of Paleontology, **78** (3): 528-541.

LATREILLE, P. A., 1802-1803. Histoire naturelle, générale et particulière, des Crustacés et des Insectes, **3**: 1-468. (F. Dufart, Paris).

— —, 1806. Genera Crustaceorum et Insectorum secundum ordinem naturalem in familias disposita, **1**: i-xviii, 1-302, pls. 1-16. (Parisiis et Argentorati).

— —, 1817a. Les Crustacés, les Arachnides et les Insectes. In: G. CUVIER, Le règne animal distribué d'après son organization, pour servir de base à l'histoire naturelle des animaux et d'introduction à l'anatomie comparée (éd. 1), **3**: i-xxix, 1-653. (Paris).

— —, 1817b. Nouveau dictionnaire d'histoire naturelle, appliquée aux arts, à l'agriculture, à l'économie rurale et domestique, à la médecine, etc., **10**: 1-404. (Déterville, Paris).

— —, 1819. Nouveau dictionnaire d'histoire naturelle appliquée aux arts, à l'agriculture, à l'économie rural et domestique, à la médecine, etc... (nouvelle édition presqu'entièrement refondue et considérablement augmentée; avec des figures tirées des troi règnes de la nature), **28**: 1-570. (Paris).

— —, 1825. Entomologie, ou histoire naturelle des Crustacés, des Arachnides et des Insectes. In: Genre de Crustacés. Encyclopédie méthodique, (Histoire naturelle), **10**: 1-832. (Paris).

— —, 1828. Encyclopédie méthodique: histoire naturelle, entomologie, ou histoire naturelle des Crustacés, des Arachnides, et des Insectes, **10**: 345-832.

— —, 1829. Les Crustacés, les Arachnides et les Insectes, distribués en familles naturelles. Le Règne Animal, distribué d'après son organization, pour servir de base à l'histoire naturelle des animaux et d'introduction à l'anatomie comparée (2nd ed.), **4**: 1-584. (Paris).

— —, 1831. Cours d'entomologie, ou de l'histoire naturelle des Crustacés, des Arachnides, des Myriapodes, et des Insectes, etc. Annales I, Atlas: 1-26. (Roret, Paris).

LAURIE, R. D., 1906. Report on the Brachyura collected by Professor Herdman, at Ceylon, in 1902. In: W. A. HERDMAN, Report to the Government of Ceylon on the Pearl Oyster Fisheries of the Gulf of Manaar with supplementary reports upon the Marine Biology of Ceylon by other Naturalists, **5** (Suppl. Rep.) (40): 349-432, pls. 1, 2.

LĂZĂRESCU, V., 1959. Asupra unei noi specii de *Coeloma* şi consideraţii paleoecologice asupra brachyurilor. Buletin Stiintific, (Secţia de Geologie şi Geografie) **2** (3-4): 665-682.

LEACH, W. E., 1814. Crustaceology. In: D. BREWSTER (ed.), Edinburgh Encyclopaedia, **7**: 383-437, pl. 221. (Edinburgh).

— —, 1814-1817. The zoological miscellany: being descriptions of new, or interesting, animals, **1-3**, 150 pls. (London).

— —, 1815-1875. Malacostraca Podophthalmata Britanniae, or descriptions of such British species of the Linnean genus *Cancer* as have their eyes elevated on footstalks: 1-124, pls. 1-45.

— —, 1816 [imprint 1815]. A tabular view of the external characters of four classes of animal which Linne arranged under Insecta: with the distribution of the genera composing three of these classes into orders and c. and descriptions of several new genera and species. Transactions of the Linnean Society of London, **11** (2): 306-400, and 'Errata'. [Dated 1815, but published 24 January 1816.]

— —, 1820. Galatéadées. Dictionnaire des Sciences Naturelles, **18**: 49-56. (F. G. Levrault, Paris).

LELOEUFF, P. & A. INTES, 1974. Les Thalassinidea (Crustacea, Decapoda) du golfe de Guinée. Systématique-Ecologie. Cahiers ORSTOM, (Océanographie) **12**: 17-69.

LEREBOULLET, A., 1858. Descriptions de deux nouvelles espèces d'écrevisses de nos rivières. Mémoires de la Société des Sciences Naturelles de Strasbourg, **5**: 1-11.

LEVITSKI, E. S., 1974. Fossil decapod crustaceans from the vicinity of the Bakhchisarai (the Crimea). Byulletin Moskovskogo Obshchestva Ispytatelei Prirody Otdel Geologicheskii, **49** (6): 101-119.

LEWIS, J. E. & A. ROSS, 1965. Notes on the Eocene Brachyura of Florida. Quarterly Journal of the Florida Academy of Sciences, **28**: 233-244.

LEWY, Z., 1977. *Ranina* (*Lophoranina*) *levantina* sp. nov. from the Middle Eocene of Israel. Israel Journal of Earth Sciences, **26**: 97-101.

LIKHAREV, B., 1917. The remains of crabs from the lower-Tertiary deposits of Donetz-Basin. Jahrbuch der Russischen Paläontologischen Gesellschaft, Petrograd, **1**: 13-24.

LIN, C.-C., 1947. On a new xanthid crab from the Neogene Formation of Taiwan (Formosa). Acta Geologica Taiwanica, **1**: 129-138.

LINNAEUS, C. [VON], 1758. Systema Naturae per Regna tria Naturae, secundum classes, ordines, genera, species, cum characteribus, differentiis, synonymis, locis (ed. 10), **1**: 1-824. (Laurentii Salvii, Holmiae [= Stockholm]).

— —, 1761. Fauna Suecica sistens Animalia Sueciae Regni: Mammalia, Aves, Amphibia, Pisces, Insecta, Vermes. Distributa per classes et ordines, genera et species, cum differentiis specierum, synonymis auctorum, nominibus incolarum, locis natalium, descriptionibus Insectorum (editio altera, auctior): 1-578. (Stockholmiae).

— —, 1767. Systema Naturae per Regna tria Naturae, secundum classes, ordines, genera, species, cum characteribus, differentiis, synonymis, locis (ed. 12), **1** (2): 533-1327.

LISSAJOUS, M., 1907. Sur un Crustacé du Bathonien Inférieure du Maconnais. Bulletin Trimestriel Société Histoire Naturelle Macon, **3**: 65-66, pl. 1.
LOCKINGTON, W. N., 1877a. Remarks on the Crustacea of the Pacific coast, with descriptions of some new species. Proceedings of the California Academy of Science, **7** [for 1876]: 28-36.
— —, 1877b. Remarks on the Crustacea of the west coast of North America, with a catalogue of the species in the Museum of the California Academy of Sciences. Proceedings of the California Academy of Science, **7** [for 1876]: 94-108.
— —, 1877c. Remarks on the Crustacea of the Pacific coast of North America, with a catalogue of the species in the Museum of the California Academy of Sciences, San Francisco. Proceedings of the California Academy of Science, **7** [for 1876]: 63-78.
LOEUFF, P. LE *see* LELOEUFF, P.
LŐRENTHEY, E., 1897. Adatok Magyarország harmadkorú rák-faunájához. Mathematikai és Természettudományi Értesito, **15**: 149-169.
— —, 1898a. Beiträge zur Decapodenfauna der ungarischen Tertiärs. Természetrajzi Füzetek, **21**: 1-133, figs. 1-9.
— —, 1898b. Über die Brachyuren der palæontologischen Sammlung des bayerischen Staates. Természetrajzi Füzetek, **21**: 134-152, pls. 10, 11.
— —, 1901. "*Andorina*" und "*Darányia*", zwei neue Brachyuren-Gattungen aus Ungarn. Mathemathische und Naturwissenschaftliche Berichte aus Ungarn, **17**: 328-336, pl. 1.
— —, 1902. Neuere Beiträge zur Tertiären Dekapodenfauna Ungarns. Mathematische und Naturwissenschaftliche Berichte aus Ungarn, **18**: 98-120, pls. 1, 2.
— —, 1907a. Palaeontologiai tanulmányok a harmadkorú rákok köréből VI. Adatok Sardinia Harmadidőszakbeli rák faunájához. Mathematikai és Természettudományi Közlemények, **29** (2): 243-295, pls. 3, 4.
— —, 1907b. Palaeontologiai tanulmányok a harmadkorú rákok köréből. V. Adatok Egyptom eocenkori rákfaunájához. Mathematikai és Természettudományi Közlemények, **29** (2): 195-242, pls. 1, 2.
— —, 1909a. Beiträge zur Tertiären Dekapodenfauna Sardiniens. Mathematische und Naturwissenschaftlichen Berichte aus Ungarn, **24**: 202-259, pls. 1, 2. [Hungarian version published in 1907.]
— —, 1909b. Beiträge zur Kenntnis der Eozänen Dekapodenfauna Ägyptens. Mathematische und Naturwissenschaftlichen Berichte aus Ungarn, **24**: 106-152, pls. 1, 2. [Hungarian version published in 1907.]
LŐRENTHEY, E. & K. BEURLEN, 1929. Die fossilen Decapoden der Länder der Ungarischen Krone. Geologica Hungarica, (Palaeontologica) **3**: 1-421, 12 tables, 16 pls.
LUND, N. T., 1793. Slaegten *Scyllarus*. Jagttagelser til Insekternes Historie, **1**. Kongelige Danske Videnskabers Selskab Skrifter, (n. ser.) **2** (2): 17-22.
MACAROVICI, N., 1970. Asupra unor fosile necunoscute din Miocenul podişului Moldovenesc. Studii şi Cercetări de Geologie-Geografie, Biologie-Muzeologie, **1**: 61-66.
MACGILCHRIST, A. C., 1905. Natural history notes from the R.I.M.S.S. "Investigator", Capt, T. H. Heming, R.N. (retired), commanding. – Series III, No. 6. An account of the new and some of the rarer decapod Crustacea obtained during the surveying seasons 1901-1904. Annals and Magazine of Natural History, **15** (87): 233-268.
MACLEAY, W. S., 1838. On the brachyurous decapod Crustacea brought from the Cape by Dr. Smith. In: A. SMITH, Illustrations of the Annulosa of South Africa; consisting chiefly of figures and descriptions of the objects of natural history collected during an expedition into the interior of South Africa, in the years 1834, 1835, and 1836; fitted out by "The Cape of Good Hope Association for Exploring Central Africa...": 53-71, 2 pls. (Smith, Elder and Company, London).

MACPHERSON, E., W. JONES & M. SEGONZAC, 2005. A new squat lobster family of Galatheoidea (Crustacea, Decapoda, Anomura) from the hydrothermal vents of the Pacific-Antarctic Ridge. Zoosystema, **27** (4): 709-723.

MAISEY, J. G., & M. G. P. DE CARVALHO, 1995. First records of fossil sergestid decapods and fossil brachyuran crab larvae (Arthropoda, Crustacea), with remarks on some supposed palaemonid fossils, from the Santana Formation (Aptian-Albian, NE Brazil). American Museum Novitates, **3132**: 1-20.

MAN, J. G. DE *see* DE MAN, J. G.

MANNING, R. B., 1987. Notes on the Western Atlantic Callianassidae (Crustacea: Decapoda: Thalassinidea). Proceedings of the Biological Society of Washington, **100** (2): 386-401.

MANNING, R. B. & F. A. CHACE, JR., 1990. Decapod and stomatopod crustaceans from Ascension Island, South Atlantic Ocean. Smithsonian Contributions to Zoology, **503**: 1-91.

MANNING, R. B. & D. L. FELDER, 1991. Revision of the American Callianassidae (Crustacea: Decapoda: Thalassinoidea). Proceedings of the Biological Society of Washington, **104**: 764-792.

MANNING, R. B. & L. B. HOLTHUIS, 1981. West African brachyuran crabs (Crustacea: Decapoda). Smithsonian Contributions to Zoology, **306**: 1-379.

— — & — —, 1989. Two new genera and nine new species of geryonid crabs (Crustacea, Decapoda, Geryonidae). Proceedings of the Biological Society of Washington, **102**: 50-77.

MANNING, R. B. & R. LEMAITRE, 1994. *Sergio*, a new genus of ghost shrimp from the Americas (Crustacea: Decapoda: Callianassidae). Nauplius, **1**: 39-43.

MANTELL, G. A., 1822. The fossils of the South Downs; or illustrations of the geology of Sussex: 1-327, 42 pls. (Lupton Relfe, London).

— —, 1833. The geology of the South-East England: 1-435. (Longman, Rees, Orme, Brown, Green and Longman, London).

— —, 1844. Medals of creation, **1-2**: 1-1016, 6 pls. (H. G. Bohn, London).

MARANGON, S. & A. DE ANGELI, 1997. *Cherpiocarcinus*, nuovo genere di brachiuro (Decapoda) dell'Oligocene del Bacino Ligure-Piemontese (Italia settentrionale). Lavori – Società Veneziana di Scienze Naturali, **22**: 97-106.

MARCK, W. VON DER *see* VON DER MARCK, W.

MARTENS, E. VON *see* VON MARTENS, E.

MARTIN, A. J., C. M. AUSTIN, L. KOOL, G. C. B. POORE, T. H. RICH, M. B. SCHULTZ & P. VICKERS-RICH, 2008. Fossil evidence in Australia for oldest known fresh-water crayfish of Gondwana. Gondwana Research, **14**: 287-296.

MARTIN, J. W. & G. B. DAVIS, 2001. An updated classification of the Recent Crustacea. Natural History Museum of Los Angeles County, Science Series, **39**: 1-124.

MARTIN, K., 1879-1880. Die Tertiärschichten auf Java. Nach den Entdeckungen von Junghuhn bearbeitet: 127-132.

MARTINS-NETO, R. G., 1987. Primeiro registro de decápode na Formação Santana, bacia do Araripe (Cretáceo Inferior), Brasil. Ciéncia e Cultura, **39** (4): 406-410.

— —, 2001. Review of some Crustacea (Isopoda and Decapoda) from Brazilian deposits (Paleozoic, Mesozoic and Cenozoic) with descriptions of new taxa. Acta Geologica Leopoldensia, **24** (52/53): 237-254.

MARTINS-NETO, R. G. & S. MEZZALIRA, 1991a. Descrição de novos crustáceos (Caridea) da Formação Santana, Cretáceo Inferior do nordeste do Brasil. Anais Academia Brasileira de Ciencias, **63** (2): 155-160.

— — & — —, 1991b. Revisão dos Palemonídeos Terciários Brasileiros (Crustacea, Caridea) com descrição de novos taxa. Anais Academia Brasileira de Ciencias, **63** (4): 361-367.

MASSALONGO, A., 1854. Geneacaena lichenum: 1-24. (Ramanzini, Verona).

MAURY, C. J., 1930. O Cretaceo da Parahyba do Norte. Ministerio da Agricultura, Industria e Commercio, Serviço Geologico e Mineralogico do Brasil, Monographia, **8**: 1-305.

MCCOY, F., 1849. On the classification of some British fossil Crustacea with notices of new forms in the university collection at Cambridge. Annals and Magazine of Natural History, (2) **4**: 161-179, 330-335.

— —, 1854. On some new Cretaceous Crustacea. Annals and Magazine of Natural History, (2) **44**: 116-122.

MCLAUGHLIN, P. A. & J. FOREST, 1997. Crustacea Decapoda: *Diacanthrus* gen. nov., a new genus of hermit crabs (Paguridae) with both Recent and fossil representation, and the descriptions of two new species. Mémoires du Muséum National d'Histoire Naturelle, Paris, **176**: 235-259.

MCLAUGHLIN, P. A. & R. LEMAITRE, 2001. A new family for a new genus and new species of hermit crab of the superfamily Paguroidea (Decapoda, Anomura) and its phylogenetic implications. Journal of Crustacean Biology, **21** (4): 1062-1077.

MCLAY, C. L., 1993. Crustacea Decapoda, the sponge crabs (Dromiidae) of New Caledonia and the Philippines with a review of the genera. In: A. CROSNIER (ed.), Résultats des campagnes MUSORSTOM, **10**. Mémoires du Muséum National d'Histoire Naturelle, Paris, **156**: 111-251.

MCLAY, C. L., R. M. FELDMANN & D. I. MACKINNON, 1995. New species of Miocene spider crabs from New Zealand, and a partial cladistic analysis of the genus *Leptomithrax* Miers, 1867 (Brachyura, Majidae). New Zealand Journal of Geology and Geophysics, **38**: 299-313.

MÉCHIN, A., 1901. Sur quelques formes nouvelles du genre *Eryma* (Astacomorpha fossiles) du Jurassique de Lorraine. Bulletin des Séances de la Societé des Sciences de Nancy, (3) **2**: 73-84.

MERTIN, H., 1941. Decapode Krebse aus dem subhercynen und Braunschweiger Emscher und Untersenon sowie Bemerkungen über verwandte Formen in der Oberkreide. Nova Acta Leopoldina, **10** (68): 149-264, pls. 1-8.

MEYER, H. VON *see* VON MEYER, H.

MICHELOTTI, G., 1861. Études sur le Miocène Inférieur de l'Italie septentrionale. (Leide [=Leyden]).

MIERS, E. J., 1876. Catalogue of the stalk- and sessile-eyed Crustacea of New Zealand: i-xii, 1-36, pls. 1-3. (Janson, London).

— —, 1879. On a collection of Crustacea made by Capt. H. C. St. John, R. N., in the Corean and Japanese Seas. Part. 1. Podophthalmia. With an appendix by Capt. H. C. St. John. Proceedings of the Zoological Society of London, **1879**: 18-61.

— —, 1886. Report on the Brachyura collected by H. M. S. Challenger during the years 1873-1876. In: C. W. THOMSON & J. MURRAY (eds.), Report of the Scientific Results of the Voyage of H. M. S. Challenger during the years 1873-1876, (Zoology): 1-362. (Johnson Reprints, New York).

MILLER, G. L. & S. R. ASH, 1988. The oldest freshwater decapod crustacean, from the Triassic of Arizona. Palaeontology, **31** (2): 273-279.

MILLET, P. A., 1854. Paléontologie de Maine et Loire: 1-187. (Cosnier et Lachèse, Angers).

MILNE-EDWARDS, A., 1860. Histoire des Crustacés podophthalmaires fossils et monographie des Décapodes macroures de la famille des Thalassiens fossiles. Annales des Sciences Naturelles, (Zoologie) (4) **14**: 129-293, pls. 1-10.

——, 1861. Études zoologiques sur les Crustacés récents de la famille des Portuniens. Archives du Muséum National d'Histoire Naturelle, Paris, **10**: 309-421, pls. 28-38.

——, 1862a. Note sur un Crustacé du terrain crétacique appartenant au genre *Porcellana*. Annales des Sciences Géologiques, **12**: [article 1].

——, 1862b. Sur l'existence de Crustacés de la famille des Raniniens pendant la période crétacée. Comptes Rendus de l'Academie des Sciences de Paris, **55**: 492-494.

——, 1862c. Monographie des Crustacés de la famille Cancériens. Annales des Sciences Naturelles, (Zoologie) (4) **18** [1862]: 31-85, pls. 1-10.

——, 1863. Monographie des Crustacés de la famille Cancériens. Annales des Sciences Naturelles, (Zoologie) (4) **20** [1863]: 273-324, pls. 5-12.

——, 1864. Monographie des Crustacés de la famille Cancériens. Annales des Sciences Naturelles, (Zoologie) (5) **1** [1864]: 31-88, pls. 1-10.

——, 1865a. Note sur deux nouveaux Crustacés fossiles du terrain néocomien du Département de l'Yonne. Bulletin de la Société des Sciences Historiques et Naturelles de l'Yonne, **19**: 341-347, pl. 5.

——, 1865b. Description de quelques Crustacés nouveaux appartenant à la tribu des Maiens. Annales de la Societé Entomologique de France, (4) **5**: 133-147.

——, 1865c. Monographie des Crustacés de la famille Cancériens. Annales des Sciences Naturelles, (Zoologie) (5) **3** [1865]: 297-351, pls. 5-13.

——, 1865d. Études zoologiques sur les Crustacés récents de la famille des Cancériens. Nouvelles Archives du Muséum d'Histoire Naturelle, Paris, **1**: 177-380, pls. 11-19.

——, 1866. Asie Mineure, description physique de celle contrée. In: P. de Tchihatcheff (ed.), Paleontologie: 165. [See A. Milne-Edwards, 1872: 4.]

——, 1867. Descriptions de quelques espèces nouvelles de Crustacés Brachyoures. Annales de la Societé Entomologique de France, **4**: 263-288.

——, 1869a. Description de quelques Crustacés nouveaux de la familie des Portuniens. Nouvelles Archives du Muséum d'Histoire Naturelle, Paris, **5**: 146-160, pls. 6, 7.

——, 1869b. Description d'un nouveau genre de Crustacé Cancérien. Annales de la Societé Entomologique de France, **9**: 167-169, pl. 8.

——, 1870. Révision du genre *Callianassa* (Leach). Nouvelles Archives du Muséum d'Histoire Naturelle, Paris, **6**: 75-101, pls. 1, 2.

——, 1872. Note sur quelques Crustacés fossiles appartenant aux genres *Ranina* et *Galenopsis*. Annales des Sciences Géologiques, **3**: 1-11.

——, 1873a. Crustacés fossiles nouveaux des terrains tertiaires de la Gironde. Actes de la Société Linnéenne de Bordeaux, **29**: 64-66.

——, 1873b. Descriptions des quelques Crustacés nouveaux ou peu connus provenant du Musée de M. C. Godeffroy. Journal des Museum Godeffroy, **1**: 77-88, 12-13.

——, 1873c. Recherches sur la faune carcinologique de la Nouvelle-Calédonie. Pt. 2. Groupe des Cyclométopes Neptunes. Nouvelles Archives du Muséum d'Histoire Naturelle, Paris, **9**: 155-332, pls. 4-18.

——, 1873-1881. Études sur les Xiphosures et les Crustacés de la région Mexicaine. Mission Scientifique au Mexique et dans l'Histoire de la Faune de l'Amérique Centrale et du Mexique, **5**: 1-368, 61 pls. (Imprimerie Nationale, Paris).

——, 1878a. Mémoire sur les Crustacés Décapodes du genre *Dynomene*. Annales des Sciences Naturelles, (Zoologie) (6) **8** (art. 3): 1-11, pls. 12-14.

——, 1878b. Description de quelques espèces nouvelles de Crustacés, provenant du Voyage aux îles du Cap Vert de MM. Bouvier et Cessac. Bulletin de la Société Philomathique de Paris, (7) **2**: 225-232.

— —, 1879. Description d'un Crustacé fossile provenant des marnes d'Aix (*Caridina nitida*). Bulletin de la Societé Philomathique de Paris, (7) **3**: 77-78.
— —, 1880a. Études préliminaires sur les Crustacés, 1ère partie. Reports on the Results of Dredging under the Supervision of Alexander Agassiz, in the Gulf of Mexico, and in the Caribbean Sea, 1877, '78, '79, by the U.S. Coast Guard Survey Steamer 'Blake', Lieutenant-Commander, C. D. Sigsbee, U.S.N., and Commander, J. R. Bartlett, U.S.N., Commanding. VIII. Bulletin of the Museum of Comparative Zoology, Harvard, **8** (1): 1-68, pls. 1, 2.
— —, 1880b. Note sur un Crustacé fossile du genre *Eumorphactaea* provenant des terrains tertiaires Inférieurs des Basses Pyrénées. Annales des Sciences Géologiques, **11**: [article 4].
— —, 1881a. Description de quelques Crustacés Macroures provenant des grandes profondeurs de la mer des Antilles. Annales des Sciences Naturelles, (Zoologie) (6) **11** (4): 1-16.
— —, 1881b. Note sur quelques Crustacés fossiles des environs de Biarritz. Annales des Sciences Géologiques, Paris, **11** (2): 1-8, pls. 21, 22.
— —, 1883. Recueil de figures de Crustacés nouveaux ou peu connus: 1-3, 44 pls. (Paris).
— —, 1886. In: A. GUILLIER, Géologie du Département de la Sarthe. (Paris et Les Mans).
MILNE-EDWARDS, A. & E.-L. BOUVIER, 1891. Observations générales sur les paguriens recueillis dans la mer des Antilles et le Golfe du Mexique, par le Blake et le Hassler, sous la direction de M. Alexandre Agassiz. Bulletin de la Société Philomatique de Paris, (8) **3**: 102-110.
MILNE-EDWARDS, A. & P. BROCCHI, 1879. Note sur quelques Crustacés fossiles appartenant au groupe des Macrophthalmiens. Bulletin de la Societé Philomathique de Paris, **3**: 113-117.
MILNE EDWARDS, H., 1830. Description des genres *Glaucothoe*, *Sicyonie*, *Sergeste* et *Acète*. Annales des Sciences Naturelles, **19**: 333-352, 4 pls.
— —, 1832. Observations sur les Crustacés du genre *Mithrax*. Magazine Zoologique, Classe 7, Crustacés, pls. 1-5.
— —, 1834-1840. Histoire naturelle des Crustacés, comprenant l'anatomie, la physiologie, et la classification de ces animaux, **1** [1834]: 1-468; **2** [1837] 1-532; **3** [1840]: 1-638, Atlas: 1-32, pls. 1-42.
— —, 1837. Description of *Dromilites*. In: L'Institut, **5**: 255.
— —, 1848. Sur quelques nouvelles espèces du genre *Pagure*. Annales des Sciences Naturelles, **10**: 59-64.
— —, 1852. De la famille des Ocypodidés (Ocypodidae). Second mémoire. In: Observations sur les affinités zoologiques et la classification naturelle des Crustacés. Annales des Sciences Naturelles, (Zoologie) **18**: 128-166, pls. 3, 4.
— —, 1853. Mémoires sur la famille des Ocypodiens, suite. Annales des Sciences Naturelles, (Zoologie) (3) **20**: 163-228, pls. 6-11.
MILNE EDWARDS, H. & H. LUCAS, 1843. Crustacés. In: A. D'ORBIGNY, Voyage dans l'Amerique méridionale (le Brésil, la république orientale de l'Uruguay, la république Argentine, la Patagonie, la république du Chili, la république de Bolivia, la république du Pérou), exécutée pendant les années 1826, 1827, 1828, 1829, 1830, 1831, 1832 et 1833, **6** (1): 1-37, pls. 1-17. (P. Bertrand, Paris).
MIYAKE, S. & H. MINEI, 1965. A new fresh-water crab *Potamon* (*Geothelphusa*) *tenuimanus* sp. nov. from Okinawa-jima, the Ryukyu Islands. Science Bulletin of the Faculty of Agriculture, Kyushu University, **21**: 377-382, pl. 21.
MOERICKE, W., 1889. Dic Crustaccen der Stramberger Schichten. Palaeontographica, (Suppl.) **2** (6): 43-72, pl. 6.

MONOD, T., 1956. Hippidea et Brachyura ouest-africains. Mémoires de l'Institut Français d'Afrique Noire, **45**: 1-674.

MONTAGU, G., 1808. Description of several marine animals found on the south coast of Devonshire. Transactions of the Linnean Society of London, **9**: 81-114.

— —, 1813. Descriptions of several new or rare animals, principally marine, discovered on the south coast of Devonshire. Transactions of the Linnaean Society of London, **11**: 1-26.

MOOSA, M. K. & R. SERÈNE, 1981. Observations on the Indo-West-Pacific Palicidae (Crustacea: Decapoda) with descriptions of two new subfamilies, four new genera and six new species. Marine Research Indonesia, **22**: 21-66.

MORET, L., 1946. *Eryma delphinensis* nouvelle espèce de Crustacé Décapode du Berriasien de Noyarey (Isère). Travaux Laboratoire de Géologie, Faculté des Sciences de Grenoble, **25**: 49-51.

MORIÈRE, J., 1864. Note sur les Crustacés du terrain jurassique du Calvados. Bulletin de la Societé Linnéenne de Normandie, **8** [1862-1863]: 89-96.

— —, 1883. Note sur une Eryonidée nouvelle trouvée à la Caine (Calvados) dans le Lias supérieur. Bulletin de la Societé Linnéenne de Normandie, (3) **7** [1882-1883]: 116-123.

— —, 1888. Note sur quelques Crustacés fossiles. Bulletin de la Societé Linnéenne de Normandie, (4) **2** [1887-1888]: 137-143.

MORRIS, S. F., 1976. A new fossil freshwater crab from the Ngorora Formation (Miocene) of Kenya. Bulletin of the British Museum (Natural History), (Geology) **27**: 295-300.

MORRIS, S. F. & J. S. H. COLLINS, 1991. Neogene crabs from Brunei, Sabah and Sarawak. Bulletin of the British Museum (Natural History), (Geology) **47** (1): 1-33.

MÜCKE, K. VON *see* VON MÜCKE, K.

MUŢIU, R. & A. BĂDĂLUŢĂ, 1971. La présence des Décapodes Anomures et Dromiacés dans les calcaires tithoniques de plate-forme Moësienne. Annales Instituti Geologici Publici Hungarici, **54** (2): 245-525, pl. 1.

MÜLLER, P., 1974a. Decapoda (Crustacea) fauna a Budapesti miocénből (2). Faune de Décapodes (Crustacés) dans le Miocène de Budapest. Földtani Közlöny, **104**: 275-287, pls. 1-3.

— —, 1974b. Decapoda (Crustacea) fauna a Budapesti miocénből. Les faunes de Crustacés Décapodes de calcaires miocènes de Budapest. Földtani Közlöny, **104**: 119-132, pls. 1-5.

— —, 1975. *Trapezia* (Crustacea Decapoda) a Magyar eocénből és miocénből. *Trapezia* (Crustacea Decapoda) dans l'Eocène et le Miocène de Hongrie. Földtani Közlöny, **105**: 516-523, pls. 1, 2.

— —, 1976. Decapoda (Crustacea) fauna a Budapesti miocénből (4). Faune de Décapodes (Crustacés) dans le Miocène de Budapest. Földtani Közlöny, **106** (2): 149-160, pls. 1-5.

— —, 1978. Decapoda (Crustacea) fauna a Budapesti miocénből (5). Faune de Décapodes (Crustacés) dans le Miocène de Budapest. Földtani Közlöny, **108**: 272-312, pls. 1-23.

— —, 1979. Crustacés Décapodes du Badenien et Sarmatien de Bulgarie. Palaeontology, Stratigraphy and Lithology, **10**: 3-7, pls. 1-3. (Sofia).

— —, 1984a. Decapod Crustacea of the Badenian. Geologica Hungarica, (Palaeontologica) **42**: 1-317, pls. 1-97.

— —, 1984b. Messinian and older decapods from the Mediterranean with description of two new species. Annales Géologiques des Pays Helléniques, **32**: 25-34, pls. 1, 2.

— —, 1993. Neogene decapod crustaceans from Catalonia. Scripta Musei Geologici Seminarii Barcinonensis, **225**: 1-39.

— —, 1996. Middle Miocene decapod Crustacea from southern Poland. Prace Muzeum Ziemi, **43**: 3-14, pls. 1, 2.

——, 1998. Decapoda Crustacea aus dem Karpat des Korneuburger Beckens (Unter-Miozän, Niederösterreich). Beiträge zur Paläontologie, **23**: 273-281.

——, 2006. New decapods from the Miocene of Hungary with remarks about their environment. Földtani Közlöny, **136** (1): 37-49.

MÜLLER, P. & J. S. H. COLLINS, 1991. Late Eocene coral-associated decapods (Crustacea) from Hungary. Contributions to Tertiary and Quaternary Geology, **28** (2-3): 47-92, pls. 1-8.

MÜLLER, P. & B. GALIL, 1998. A note on a Miocene matutine crab from Hungary (Decapoda, Brachyura, Calappidae). Crustaceana, **71** (5): 583-588.

MÜNSTER, G. G. ZU *see* ZU MÜNSTER, G. G.

NAGAO, T., 1931. Two new decapod species from the Upper Cretaceous deposits of Hokkaido, Japan. Journal of the Faculty of Science, Hokkaidô Imperial University, (4, Geology and Mineralogy) **1** (2): 207-214, pls. 1-14.

——, 1932. Two Tertiary and one Cretaceous Crustacea from Hokkaidô, Japan. Journal of the Faculty of Science, Hokkaidô Imperial University, (4, Geology and Mineralogy) **2** (1): 15-22, pl. 4.

——, 1940. On a new Brachyura crab from the Ttatunokuti bed of Sendai, Miyagi Prefecture. Journal of the Faculty of Science, Hokkaidô Imperial University, (4, Geology and Mineralogy) **6** (1): 69-73, pl. 23.

——, 1941. On some fossil Crustacea from Japan. Journal of the Faculty of Science, Hokkaido Imperial University, (4, Geology and Mineralogy) **6**: 85-100, pl. 26.

NAGAO, T. & K. HUZIOKA, 1938. A new species of *Callianassa* from the Neogene Tertiary of Hokkaido. Journal of the Faculty of Science, Hokkaidô Imperial University, (4, Geology and Mineralogy) **4** (1, 2): 63-67, pl. 6.

NAGAO, T. & K. ÔTATUME, 1938. A new *Callianassa* from the Paleogene Isikari Series of Hokkaido. Journal of the Geological Society of Japan, **45** (540): 101-102.

NARDO, G. D., 1847. Sinonimia moderna delle specie registrate nell'opera intitolata: descrizione de'Crostacei, de'Pesci che abitano le lagune e gulfo Veneto rappresentari in fugure, a chiaro-scuro ed a colori dell'Abate Stefano Chiereghini Ven. Clodiense applicata per commissione governativa: i-xi, 1-127. (Venice).

NATIONS, D. J., 1968. A new species of cancroid crab from the Pliocene of California. Journal of Paleontology, **42**: 33-36.

——, 1975. The genus *Cancer* (Crustacea: Brachyura): systematics, biogeography and fossil record. Natural History Museum of Los Angeles County Science Bulletin, **23**: 1-104, pls. 1-80.

NG, N. K., P. J. F. DAVIE, C. D. SCHUBART & P. K. L. NG, 2007. Xenograpsidae, a new family of grapsoid crabs (Crustacea: Brachyura) associated with shallow water hydrothermal vents. Raffles Bulletin of Zoology, (Suppl.) **16**: 233-256.

NG, P. K. L., 1993. Kraussiinae, a new subfamily for the genera *Kraussia* Dana, 1852, *Palapedia*, new genus, and *Garthasia*, new genus (Crustacea: Decapoda: Brachyura: Xanthidae), with descriptions of two new species from Singapore and the Philippines. Raffles Bulletin of Zoology, **41** (1): 133-157.

NG, P. K. L. & D. G. B. CHIA, 1994. The genus *Glyptocarcinus* Takeda, 1973, with descriptions of a new subfamily, two new genera and two new species from New Caledonia (Crustacea: Decapoda: Brachyura: Xanthidae). Raffles Bulletin of Zoology, **42** (3): 701-730.

NG, P. K. L. & P. F. CLARK, 2000. The Indo-Pacific Pilumnidae XII. On the familial placement of *Chlorodiella bidentata* (Nobili, 1901) and *Tanaocheles stenochilus* Kropp, 1984 using adult and larval characters with the establishment of a new subfamily, Tanaochelinae (Crustacea: Decapoda: Brachyura). Journal of Natural History, London, **34**: 207-245.

NG, P. K. L., D. GUINOT & P. J. F. DAVIE, 2008. Systema Brachyurorum: Part I. An annotated checklist of extant brachyuran crabs of the world. Raffles Bulletin of Zoology, (Suppl.) **17**: 1-286.
NG, P. K. L. & L. B. HOLTHUIS, 2007. *Etisus* H. Milne Edwards, 1834 and *Chlorodiella* Rathbun, 1897 (Crustacea, Decapoda, Brachyura): proposed conservation of the generic names by suppression of the generic name *Clorodius* A. G. Desmarest, 1823. Bulletin of Zoological Nomenclature, **64**: 1 [case 3394 (Opinion 2204)].
NG, P. K. L. & M. R. MANUEL-SANTOS, 2007. Establishment of the Vultocinidae, a new family for an unusual new genus and new species of Indo-West Pacific crab (Crustacea: Decapoda: Brachyura: Goneplacoidea), with comments on the taxonomy of the Goneplacidae. Zootaxa, **1558**: 39-68.
NG, P. K. L. & G. RODRIGUEZ, 1986. New records of *Mimilambrus wileyi* Williams, 1979 (Crustacea: Decapoda: Brachyura), with notes on the systematics of the Mimilambridae Williams, 1979 and Parthenopoidea MacLeay, 1838 sensu Guinot, 1978. Proceedings of the Biological Society of Washington, **99**: 88-99.
NOETLING, F., 1881. Ueber einige Brachyuren aus dem Senon von Maestricht und dem Tertiär Norddeutschlands. Zeitschrift der Deutschen Geologischen Gesellschaft, **33**: 357-371, pl. 20.
— —, 1885a. Die Fauna des samländischen Tertiärs. Abhandlungen zur Geologischen Spezialkarte von Preussen, **6** (2) (1) (2): 112-172.
— —, 1885b. Über Crustaceen aus dem Tertiär Aegyptens. Sitzungsberichte der Königlichen Preussischen Akademie der Wissenschaften, **26**: 487-500, pl. 4.
— —, 1897. Fauna of the Upper Cretaceous (Maëstrichtien) beds of the Mari Hills. Memoirs of the Geological Survey of India, Palaeontologica Indica, (16) **1** (3): 1-79, pls. 1-23.
— —, 1901. Fauna of the Miocene beds of Burma. Memoirs of the Geological Survey of India, Palaeontologica Indica, (new series) **1**: 1-378, 25 pls.
NYBORG, T. G., R. E. BERGLUND & J. L. GOEDERT, 2003. A new crab from the late Eocene Hoko River Formation, Olympic Peninsula, Washington: the earliest record of *Euphylax* (Decapoda: Portunidae). Journal of Paleontology, **77** (2): 323-330.
NYBORG, T. G. & J. FAM, 2008. *Bicornisranina bocki*, n. gen., n. sp. (Decapoda: Raninidae) from the Cretaceous of Vancouver Island, British Columbia, Canada. Journal of Crustacean Biology, **28**: 686-694.
NYBORG, T. G. & F. J. VEGA, 2008. Three new fossil species of *Lophomastix* (Decapoda: Blepharipodidae) from the Cenozoic of Washington. Journal of Crustacean Biology, **28**: 361-369.
ODHNER, T., 1925. Monographierte Gattungen der Krabben-Familie Xanthidae. I. Göteborgs Kungliga Vetenskaps- och Vitterhets-Samhalles Handlingar, **29** (1): 1-92, pls. 1-5.
OLIVI, G., 1792. Zoologia Adriatica: ossia catalogo ragionato degli animali del Golfo e delle Lagune di Venezia; preceduto da una dissertazione sull storia fisica e naturale del Golfo; e accompagnato da memorie, ed osservazioni di fisica storia naturale ed economia: 1-334, 9 pls. (G. Remondini, Bassano).
OPPEL, A., 1854. Der Mittlere Lias Schwabens. Jahreshefte des Vereins für Vaterländische Naturkunde in Württemberg, **10**: 39-136.
— —, 1861a. Die Arten der Gattungen *Glyphea* und *Pseudoglyphea*. Jahreshefte des Vereins für Vaterländische Naturkunde in Württemberg, **17**: 108-111.
— —, 1861b. Die Arten der Gattungen *Eryma*, *Pseudastacus*, *Magila* und *Etallonia*. Jahreshefte des Vereins für Vaterländische Naturkunde in Württemberg, **17**: 355-361.
— —, 1862. Ueber jurassische Crustaceen. Palaeontologische Mitteilungen aus dem Museum der K. Bayerischen Staates, **1**: 1-120.

OPPENHEIM, P., 1899. I supposti rapporti dei crostacei terziarii di Ofen descritti da Loerenthey con quelli veneti. Rivista Italiana di Paleontologia, **5**: 55-62.

— —, 1901. Die Priabonaschichten und ihre Fauna. Palaeontographica, **47**.

ORAVECZ, J., 1962. Der erste Macrurenfund *Paraclytiopsis hungaricus* nov. gen. nov. sp. aus dem ungarischen Karn. Földtani Közlöny, **92** (3): 324-329, pl. 45.

ORBIGNY, A. D' *see* D'ORBIGNY, A.

ORDWAY, A., 1863. Monograph of the genus *Callinectes*. Journal of the Boston Society of Natural History, **7**: 567-583.

ORTMANN, A. E., 1890. Die Unterordnung Natantia Boas: die Decapoden-Krebse des Strassburger Museums mit besonderer Berücksichtigung der von Herrn Dr. Döderlein bei Japan und bei den Liu-Kiu-Inseln gesammelten und z. Z. im Strassburger Museum aufbewahrten Formen, I. Zoologische Jahrbücher, (Systematik, Ökologie und Geographie der Tiere) **5**: 437-542, pls. 36, 37.

ORTMANN, A. [E.], 1891. Die Decapoden-Krebse des Strassburger Museums. 3. Die Abtheilungen der Reptantia Boas: Homaridae, Loricata und Thalassinidea. Zoologischen Jahrbücher, (Systematik, Ökologie, und Geographie der Tiere) **6**: 1-58.

— —, 1892a. Die Dekapoden-Krebse des Strassburger Museums IV. Die Abtheilungen Galatheidea und Paguridea. Zoologischen Jahrbücher, (Systematik, Geographie und Biologie der Tiere) **6**: 241-326, pls. 11, 12.

— —, 1892b. Die Abtheilungen Hippidea, Dromiidea und Oxystomata: die Decapoden-Krebse des Strassburger Museums, mit besonderer Berücksichtigung der von Herrn Dr. Döderlein bei Japan und bei den Liu-Kiu-Inseln gesammelten und z. Z. im Strassburger Museum aufbewahrten Formen. V Theil. Zoologische Jahrbücher, (Systematik, Geographie und Biologie der Thiere) **6**: 532-588, pl. 26.

— —, 1893. Abtheilung: Brachyura (Brachyura genuina Boas), II. Unterabtheilung: Cancroidea, 2. Section: Cancrinea, 1. Gruppe: Cyclometopa. Die Decapoden-Krebse des Strassburger Museums, mit besonderer Berücksichtigung der von Herrn Dr. Döderlein bei Japan und bei den Liu-Kiu-Inseln gesammelten und zur Zeit im Strassburger Museum aufbewahrten Formen, VII Theil. (Zoologische Jahrbücher, (Systematik, Geographie und Biologie der Thiere) **7**: 411-495, pl. 17.

— —, 1894. Die Decapoden-Krebse des Strasburger Museums. Zoologische Jahrbücher, (Systematik) **8**: 683-772.

— —, 1895. A study of the systematic and geographic distribution of the decapod family Crangonidae Bate. Proceedings of the Academy of Natural Science of Philadelphia, **47**: 173-197.

— —, 1896. Das System der Decapoden-Krebse. Zoologische Jahrbücher, (Systematik, Geographie und Biologie der Thiere) **9**: 409-453.

— —, 1897. On a new species of the palinurid-genus *Linuparus* found in the Upper Cretaceous of Dakota. American Journal of Science, (4) **4**: 290-297.

— —, 1898. Gliederfüssler: Arthropoda. In: Dr. H. G. BRONN'S, Klassen und Ordnungen des Tierreichs, **5** (2), Crustacea, Malacostraca, (47-52) [Systematik]: 1057-1168.

— —, 1905. The mutual affinities of the species of the genus *Cambarus*, and their dispersal over the United States. Proceedings of the American Philosphical Society, **44** (180): 91-136, pl. 3.

PACKARD, A. S., 1880. Fossil crawfish from the tertiaries of Wyoming. American Naturalist, **14**: 222-223.

— —, 1900. A new fossil crab from the Miocene Greensand-bed of Gayhead, Martha's Vineyard with remarks on the phylogeny of the genus *Cancer*. Proceedings of the American Academy of Arts and Sciences, **36**: 1-9, 2 pls.

PARKER, T. J., 1883. On the structure of the head in *Palinurus* with special reference to the classification of the genus. Nature, London, **29**: 189-190, 1 pl.
PATEK, S. N., R. M. FELDMANN, M. PORTER & D. TSHUDY, 2006. Phylogeny and evolution. In: B. F. PHILLIPS (ed.), Lobsters: biology, management, aquaculture and fisheries: 113-146. (Blackwell Publishing, Oxford, U.K.).
PATRULIUS, D., 1959. Contributions à la systématique des Décapodes néojurassiques. Revue de Géologie et Géographie, **3** (2): 249-257.
——, 1966. Les Décapodes du Tithonique Inférieur de Woźniki (Carpates Polonaises Occidentales). Annales de la Société Géologique de Pologne, **36** (4): 495-517, pls. 30, 31.
PĂUCĂ, M., 1929. Zwei neue fossile Decapoden aus den Oligozänen *Clupea crenata* Schiefern von Susläneski-Muscel und über die Bildung der Menilitschiefer. Bulletin de la Section Scientifique de l' Académie Roumaine, **12** (4-5): 40-44.
PAUL'SON, O., 1875. Studies on the Crustacea of the Red Sea with notes regarding other seas. Part 1. Podophthalmata and Edriophthalmata (Cumacea): 1-164, 21 pls. (S. V. Kul'zhenko, Kiev). [Translation published for The National Science Foundation, Washington, D.C. and Smithsonian Institution, U.S.A., by the Israel Program for Scientific Translations, 1961.]
PELSENEER, P., 1886a. Notice sur un Crustacé des Sables verts de Grandpré. Bulletin du Musée Royal d'Histoire Naturelle de Belgique, **4** (2): 47-59.
——, 1886b. Notice sur un Crustacé de la Craie brune des environs de Mons. Bulletin du Musée Royal d'Histoire Naturelle de Belgique, **4** (1): 37-46, pls. 1-3.
——, 1886c. Notice sur les Crustacés Décapodes du Maastrichtien du Limbourg. Bulletin du Musée Royal d'Histoire Naturelle de Belgique, Bruxelles, **4** (3): 161-175.
PENNANT, T., 1777. British Zoology, **4**. Crustacea, Mollusca, Testacea. (Benjamin White, London).
PETERS, W., 1861. Eine neue Eintheilung der Skorpione und über die von ihm in Mossambique gesammelten Arten von Skorpionen, aus welchem hier Auszug mitgetheilt wird. Monatsberichte der Königlichen Preussischen Akademie der Wissenschaften zu Berlin, **1861**: 507-516.
PHILIPPE, M. & S. SECRETAN, 1971. Crustacés Décapodes du Burdigalien des Courennes (Vaucluse). Annales de Paléontologie, (Invertébrés) **57**: 117-134.
PHILIPPI, R. A., 1838. *Palicus granulatus*, ein neues Genus der rückenfüssigen Krabben. Jahresberichte des Vereins für Naturkunde zu Kassel, **2**: 11-12.
——, 1887. Los fósiles terciarios I cuartarios de Chile: 1-256, 56 pls. (Brockhaus, Leipzig [German version] and Santiago de Chile [Spanish version]).
PHILLIPS, J., 1829. Illustrations of the geology of Yorkshire, Part 1. The Yorkshire coast: 1-184. (John Murray, London).
——, 1836. Illustrations of the geology of Yorkshire (2nd ed.). (London).
——, 1871. Geology of Oxford and the Valley Thames: 1-523. (Clarendon Press, Oxford).
PILSBRY, H. A., 1901. Crustacea of the Cretaceous Formation of New Jersey. Proceedings of the Academy of Natural Sciences of Philadelphia, **53**: 111-118.
——, 1916. Systematic paleontology of the Upper Cretaceous deposits of Maryland (Arthropoda). Maryland Geological Survey, (Upper Cretaceous) **1916**: 361-370, pls. 10, 11.
PINNA, G., 1968. Gli Erionidei della nuova fauna sinemuriana a crostacei decapodi di Osteno in Lombardia. Atti della Società italiana di Scienze naturali e del Museo civico di Storia naturale in Milano, **107**: 93-134.
——, 1974. I crostacei della fauna triassica di Cene in Val Seriana (Bergamo). Memorie della Società italiana di Scienze naturali e del Museo civico di Storia naturale in Milano, **21** (1): 5-34.

PLAZIAT, J.-C. & S. SECRETAN, 1971. La faune de Crustacés Décapodes des calcaires à Alvéolines yprésiens des Corbbières septentrionales (Aude). Géobios, **4** (2): 117-142, pls. B, 8, 9.

POEPPIG, E., 1836. Crustacea Chilensia nova aut minus nota descripsit. Archiv für Naturgeschichte, **2** (1): 133-145.

POLKOWSKY, S., 2004. Decapode Krebse aus dem Oberoligozenen Sternberger Gestein von Kobrow (Mecklenburg). Tassados, **1**: 1-126.

POLZ, H., 1999. *Etallonia hoellorum* sp. nov. (Crustacea, Decapoda, Axiidae) aus dem Oberkimmeridgium der südlichen Frankenalb. Archaeopteryx, **17**: 33-39.

— —, 2000. *Glyphea viohli* sp. nov. (Crustacea: Decapoda: Glypheidae) aus den Solnhofener Plattenkalken. Archaeopteryx, **18**: 17-25.

— —, 2007. Die Garnelengattung *Harthofia* g. nov. (Crustacea: Decapoda: Pleocyemata: Caridea) mit zwei neuen Arten aus den Solnhofener Plattenkalken von Eichstätt. Archaeopteryx, **25**: 1-13.

— —, 2009 [imprint 2008]. *Almonacaris winkleri* g. nov. sp. nov. (Crustacea: Decapoda: Pleocyemata: Caridea) aus den Solnhofener Plattenkalken von Eichstätt. Archaeopteryx, **26**: 1-9.

POORE, G. C. B., 1994. A phylogeny of the families of Thalassinidea (Crustacea: Decapoda) with keys to families and genera. Memoirs of the Museum of Victoria, **54**: 79-120.

PORTELL, R. W. & J. S. H. COLLINS, 2002. A new species of *Montezumella* (Crustacea, Decapoda, Cheiragonidae) from the Upper Eocene Ocala Limestone of Florida. Proceedings of the Biological Society of Washington, **115** (3): 594-599.

— — & — —, 2004. Decapod crustaceans of the Lower Miocene Montpelier Formation, White Limestone Group of Jamaica. Cainozoic Research, **3** (1-2): 109-126.

POWELL, A. W. B., 1949. New species of Crustacea from New Zealand of the genera *Scyllarus* and *Ctenocheles*, with notes on *Lyreidus fastigatus*. Records of the Auckland Institute Museum, **3**: 368-371, 1 pl.

PRASAD, K. N., 1966. Crustacea from the Fuller's Earth deposits of Kapurdi, Rajasthan. Records of the Geological Survey of India, **94**: 313-316.

PRESTANDREA, N., 1833. Su di alcuni nuovi Crustacei dei mari di Messina. Effemeridi Scientifiche per La Sicilia, **1833** (April): 3-14.

— —, 1839. Descrizione di due nuovi Crustacei dei mari di Messina. Atti dell'Accademia Gioenia di Scienze Naturali di Catania, (1) **14**: 131-136.

PRETZMANN, G., 1968. Die Familie Trichodactylidae (Milne Edwards, 1853) Smith (vorläufige Mitteilung). Entomologisches Nachrichtenblatt, **15**: 70-76. (Vienna).

— —, 1978. Neue Süßwasserkrabben aus den Anden. Sitzungsberichten der Österreichischen Akademie der Wissenschaften, (Mathematisch-naturwissenschaftliche Klasse) (1) **187**: (6-10), 163-170.

PUSCH, A., 1838. Ueber einen fossilen Krebs aus Polen (*Astacus leucoderma*). Neues Jahrbuch für Mineralogie, Geologie, Geognosie, und Petrefaktenkunde, **1838**: 130-135.

QUAYLE, W. J., 1984. A new crab, *Portunites stintoni* (Crustacea, Decapoda) from the London Clay. Tertiary Research, **5**: 173-178.

— —, 1987. English Eocene Crustacea (lobsters and stomatopod). Palaeontology, **30**: 581-612.

QUAYLE, W. J. & J. S. H. COLLINS, 1981. New Eocene crabs from the Hampshire Basin. Palaeontology, **24** (4): 733-758, pls. 104, 105.

QUENSTEDT, F. A., 1851. Handbuch der Petrefaktenkunde: 1-792, Atlas. (H. Laupp, Tübingen).

— —, 1854 [imprint 1850]. Ueber *Mecochirus* im braunen Jura bei Gammelshausen und einige andere Krebse. Jahreshefte des Vereins für Vaterländische Naturkunde in Württemberg, **6**: 186-197.

— —, 1856-1858. Der Jura: 1-842, 100 pls. (Verlag der H. Lauppschen Buchhandlung, Tübingen).

— —, 1867. Handbuch der Petrefaktenkunde: 1-982, Atlas with 86 pls. (H. Laupp, Tübingen).

RABADÀ, D., 1990. Crustáceos decápodos de Las Hoyas (Cuenca) y del Montsec de Rubies (Lleida). Calizas Litograficas del Cretacico Inferior de España. Acta Geológica Hispanica, Madrid, **25**: 299-311.

— —, 1993. Crustáceos decápodos lacustres de las calizas litograficas del Cretacico Inferior de España: Las Hoyas (Cuenca) y el Montsec de Rubies (Lleida). Cuadernos de Geologìa Ibérica, Madrid, **17**: 345-370.

RAFINESQUE, C. S., 1814. Précis des découvertes et travaux somiologiques entre 1800 et 1814 ou choix raisonné de ses principales découvertes en zoologie et en botanique, pour servir d'introduction à ses ouvrages futurs: 1-56. (C. S. Rafinesque-Schmaltz, Palermo).

— —, 1815. Analyse de la nature, ou tableau de l'univers et des corps organisée: 1-224. (L'Imprimerie de Jean Barravecchia, Palermo).

RANDALL, J. W., 1839. Catalogue of the Crustacea brought by Thomas Nuttall and J. K. Townsend, from the west coast of North America and the Sandwich Islands, with descriptions of such species as are apparently new, among which are included species of different localities, previously existing in the collection of the Academy. Journal of the Academy of Natural Sciences of Philadelphia, **8**: 106-147.

RANZANI, C., 1820. Osservazioni sopra un fossile dell'Aldrovandi chiamato Sepite. Memorie di Storia Naturale, (Decade Prima) **7**. (Bologna).

RATHBUN, M. J., 1892. Catalogue of the crabs of the family Periceridae in the U.S. National Museum. Proceedings of the United States National Museum, **15**: 231-277, pls. 28-40.

— —, 1894. Descriptions of two new species of crabs from the western Indian Ocean, presented to the National Museum by Dr. W. L. Abbott. Proceedings of the United States National Museum, **17** (979): 21-24.

— —, 1896. The genus *Callinectes*. Proceedings of the United States National Museum, **18** (1070): 349-375, pls. 12-28.

— —, 1897a. A revision of the nomenclature of the Brachyura. Proceedings of the Biological Society of Washington, **11**: 153-167.

— —, 1897b. Description of a new species of *Cancer* from Lower California, and additional note on *Sesarma*. Proceedings of the Biological Society of Washington, **11**: 111-112.

— —, 1897c. List of the decapod Crustacea of Jamaica. Annals of the Institute of Jamaica, **1** (1): 3-46.

— —, 1898a. The Brachyura collected by the U.S. Fish Commission steamer Albatross on the voyage from Norfolk, Virginia, to San Francisco, California, 1887-1888. Proceedings of the United States National Museum, **21**: 567-616, pls. 41-44.

— —, 1898b. The Brachyura of the Florida Keys and the Bahamas in 1893. Bulletin from the Laboratories of Natural History of the State University of Iowa, **4** (3): 349-375.

— —, 1898c. A contribution to a knowledge of the fresh-water crabs of America. – The Pseudothelphusinae. Proceedings of the United States National Museum, **21**: 507-537.

— —, 1900. Synopses of North-American invertebrates. VII. Cyclometopous or cancroid crabs of North America. The American Naturalist, **34**: 131-143.

— —, 1902a. Descriptions of new decapod crustaceans from the west coast of North America. Proceedings of the U.S. National Museum, **24**: 885-905.

— —, 1902b. Papers from the Hopkins Stanford Galapagos expedition 1898-1899. VIII. Brachyura and Macrura. Proceedings of the Washington Academy of Sciences, **4**: 275-292.
— —, 1904a. Some changes in Crustacean nomenclature. Proceedings of the Biological Society of Washington, **17**: 169-172.
— —, 1904b. Les Crabes d'eau douce (Potamonidae). Nouvelles Archives du Muséum d'Histoire Naturelle, Paris, (4) **6**: 225-312, pls. 9-18.
— —, 1906. The Brachyura and Macrura of the Hawaiian Islands. U.S. Fish Commission Bulletin for **1903** (3): 829-930, pls. 3-24.
— —, 1908. Descriptions of fossil crabs from California. Proceedings of the United States National Museum, **35**: 341-349.
— —, 1909. New crabs from the Gulf of Siam. Proceedings of the Biological Society of Washington, **22**: 107-114.
— —, 1914. A new genus and some new species of crabs of the family Goneplacidae. Proceedings of the United States National Museum, **48**: 137-154.
— —, 1915. *Jacquinotia*, a new crab name. Proceedings of the Biological Society of Washington, **28**: 142.
— —, 1916. Description of a new genus and species of fossil crab from Port Townsend, Washington. American Journal of Science, **41**: 344-346.
— —, 1917a. New species of South Dakota Cretaceous crabs. Contributions to Zoology, **67**: 237-255.
— —, 1917b. Description of a new crab from the California Pliocene. Proceedings of the United States National Museum, **53**: 451-452.
— —, 1918. Decapod crustaceans from Panama. In: T. W. VAUGHAN (ed.), Contributions to the geology and paleontology of the Canal Zone, Panama and geologically related areas in Central America and the West Indies. United States National Museum Bulletin, **103**: 123-184, pls. 54-66.
— —, 1919. West Indian Tertiary decapod crustaceans. In: T. W. VAUGHAN (ed.), Contributions to the geology and paleontology of the West Indies: 159-184, pls. 1-9. (Carnegie Institution of Washington Publication, **291**, Washington, D.C.).
— —, 1920a. Additions to the West Indian Tertiary decapod crustaceans. Proceedings of the United States National Museum, **58**: 381-384, pl. 25.
— —, 1920b. New species of spider crabs from the Straits of Florida and Caribbean Sea. Proceedings of the Biological Society of Washington, **33**: 23-24.
— —, 1923a. Decapod crustaceans from the Upper Cretaceous of North Carolina. North Carolina Geological Survey, **5**: 403-407.
— —, 1923b. Fossil crabs from the Republic of Haiti. Proceedings of the Biological Society of Washington, **63**: 1-6, pls. 1, 2.
— —, 1923c. Descriptions of new species of crabs from Palmyra Island. The Bishop Museum Bulletin, **5**: 38-40.
— —, 1925. The spider crabs of America. United States National Museum Bulletin, **129**: 1-613.
— —, 1926a. Crustacea. In: B. WADE (ed.), The fauna of the Ripley Formation of Coon Creek, Tennessee. U.S. Geological Survey Professional Paper, **137**: 184-191, pls. 63-70.
— —, 1926b. The fossil stalk-eyed Crustacea of the Pacific slope of North America. United States National Museum Bulletin, **138**: i-viii, 1-155.
— —, 1928. Two new crabs from the Eocene of Texas. Proceedings of the United States National Museum, **73**: 1-6, pls. 1-3.
— —, 1929a. New species of fossil decapod crustaceans from California. Journal of the Washington Academy of Sciences, **19** (21): 469-472.

——, 1929b. A new crab from the Eocene of Florida. Proceedings of the United States National Museum, **75**: 1-4, pls. 1-3.
——, 1929c. New and rare Chinese crabs. Lingnan Science Journal, **8**: 75-104, pls. 5-15.
——, 1930a. A new *Callianassa* from the Cretaceous of South Dakota. Journal of the Washington Academy of Sciences, **20**: 1-3.
——, 1930b. Fossil decapod crustaceans from Mexico. Proceedings of the United States National Museum, **78**: 1-10, pls. 1-6.
——, 1930c. The cancroid crabs of America of the families Euryalidae, Portunidae, Atelecyclidae, Cancridae and Xanthidae. United States National Museum Bulletin, **152**: 1-609.
——, 1931a. A new fossil palinurid from Staten Island. Proceedings of the Staten Island Institute of Arts and Sciences, **5** (2-4): 161-162.
——, 1931b. Two new crabs from the Gulf of Mexico. Proceedings of the Biological Society of Washington, **44**: 71-72.
——, 1932a. New species of fossil Raninidae from Oregon. Journal of the Washington Academy of Science, **22**: 239-242.
——, 1932b. A new species of *Cancer* from the Pliocene of the Los Angeles basin. Journal of the Washington Academy of Sciences, **22** (1): 1.
——, 1932c. Fossil pinnotherids from the California Miocene. Journal of the Washington Academy of Sciences, **22**: 411-413.
——, 1934. Fossil decapod crustaceans from Vitilevu, Fiji. Bernice P. Bishop Museum Bulletin, **119**: 238-241, pl. 44.
——, 1935a. Fossil Crustacea of the Atlantic and Gulf Coastal Plain. Geological Society of America, (special paper) **2**: i-viii, 1-160.
——, 1935b. Preliminary descriptions of seven new species of oxystomatous and allied crabs. Proceedings of the Biological Society of Washington, **48**: 1-4.
——, 1935c. A new xanthid crab from the Cretaceous of New Jersey. Proceedings of the Academy of Natural Science, Philadelphia, **87**: 165-166.
——, 1936. Corrections of names of fossil decapod crustaceans. Proceedings of the Biological Society of Washington, **49**: 37.
——, 1937. Cretaceous and Tertiary crabs from Panama and Colombia. Journal of Paleontology, **11**: 26-28, pl. 5.
——, 1945. Decapod Crustacea. In: H. S. LADD & J. E. HOFFMEISTER (eds.), Geology of Lau, Fiji. Bernice P. Bishop Museum Bulletin, **181**: 373-391. (Honolulu).
——, 1947. Phylum Arthropoda. In: M. M. KNECHTEL, E. F. RICHARDS & M. J. RATHBUN (eds.), Mesozoic fossils of the Peruvian Andes. Johns Hopkins University Studies in Geology, **15**: 1-150.
RAVN, J. P. J., 1903. The Tertiary fauna at Kap Dalton (Grönland). Meddelelser om Grønland, **29**: 93-140.
REED, F. R. C., 1911. New Crustacea from the Lower Greensand of the Isle of Wight. Geological Magazine, **8**: 115-120.
REMEŠ, M., 1895. Beiträge zur Kenntnis der Crustaceen der Stramberger Schichten. Bulletin International de l'Académie des Sciences de Bohème, **2**: 200-204, pls. 1-3.
REMY, J.-M., 1960. Études paléontologiques et géologiques sur les Falaises de Fresco (Côte d'Ivoire) 2. Crustacés. Annales de la Faculté des Sciences, Université de Dakar, **5**: 55-64.
REMY, J. M. & F. TESSIER, 1954. Décapodes nouveaux de la partie ouest du Sénégal. Bulletin de la Société Géologique de France, **6** (4): 185-191, pl. 11.
RENAULT, CH., 1889. Note sur une Eryonidée nouvelle trouvée à St. Honorine-la-Guillaume (Orne) dans le grès liasique. Bulletin de la Societé Linnéenne de Normandie, (4) **2**: 13-19.
REUSS, A. E., 1845. Die Versteinerungen der böhmischen Kreideformation. (Stuttgart).

— —, 1857. Zur Kenntnis fossiler Krabben. Sitzungsberichte der Akademie der Wissenschaften, **27**: 161-166.

— —, 1858 [imprint 1857]. Über kurzschwänzige Krebse im Jurakalke Mährens. Sitzungsberichte der Kaiserlichen Akademie der Wissenschaften, (Mathematisch-naturwissenschaftliche Classe) **31**: 5-13.

— —, 1858. Ueber fossile Krebse aus den Raibler Schichten in Kaernthen. Beiträge zur Paläontographie Oesterreichs, **1**: 1-6.

— —, 1859. Zur Kenntnis fossiler Krabben. Akademie der Wissenschaften Wien, Denkschrift, **17**: 1-90, pls. 1-24.

— —, 1867. Die fossile Fauna der Steinsalzablagerungen von Wieliczka in Galizien. Sitzungsberichte der Akademie der Wissenschaften in Wien, (Mathematisch-naturwissenschaftliche Klasse) **40** (1): 173-176, pl. 8.

REUSS, R. VON *see* VON REUSS, R.

RICHARDS, B. C., 1975. *Longusorbis cuniculosus*: a new genus and species of Upper Cretaceous crab with comments on the Spray Formation at Shelter Point, Vancouver Island, British Columbia. Canadian Journal of Earth Sciences, **12**: 1850-1863.

RISSO, A., 1815. Histoire naturelle des Crustacés des environs de Nice: 1-175, pls. 1-3. (Paris).

— —, 1826. Histoire naturelle des principales productions de l'Europe méridionale et particulièrement de celles des environs de Nice et des Alpes Maritimes, **4**: [1-3], j-vij [= 1-7], 1-439, pls. [1-12]. (Levrault, Paris).

RISTORI, G., 1886. I crostacei brachiuri e anomuri del pliocene italiano. Bollettino della Società Geologica Italiana, **5**: 93-128, pls. 2, 3.

— —, 1888. Alcuni crostacei del miocene medio italiano. Atti della Società Toscana di Scienze Naturali (Pisa), **9**: 212-219, pls. 2, 3.

— —, 1889a. Crostacei Piemontesi del Miocene Inferiore. Bollettino della Società Geologica Italiana, **7**: 397-412.

— —, 1889b. Un nuovo crostaceo fossile del Giappone. Atti della Società Toscana di Scienze Naturali (Pisa), [verbal proceedings] **7**: 4-6.

— —, 1891. Contributo alla fauna carcinologica del Pliocene italiano. Atti della Società Toscana di Scienze Naturali (Pisa), **11**: 1-18, pl. 1.

— —, 1893. Note di carcinologia pliocenica. Atti della Società Toscana di Scienze Naturali residente in Pisa, Processi Verbali, **8**: 86-89.

— —, 1896. Crostacei neogenici di Sardegna e di alcune altre localita italiane. Bollettino della Società Geologica Italiana, **15**: 504-513, pl. 12.

ROBERTS, H. B., 1953. A new decapod Crustacean from the Inglis Member. In: Eocene mollusks from Citrus and Levy counties. Geological Bulletin, **35** (29): 64-67.

— —, 1955. New xanthid crab from the Claiborne Eocene of New Jersey. Bulletin of the Wagner Free Institute of Science, **30** (1): 9-12, pl. 1.

— —, 1956. Early Tertiary decapod crustaceans from the Vincentown Formation in New Jersey. Bulletin of the Wagner Free Institute of Science, **31** (2): 5-12, pl. 2.

— —, 1962. The Upper Cretaceous decapod crustaceans of New Jersey and Delaware. In: H. G. RICHARDS (ed.), The Cretaceous fossils of New Jersey. Bulletin of the New Jersey Division of Geology, **61**: 163-192.

ROBINEAU-DESVOIDY, J. B., 1849. Mémoire sur les Crustacés du terrain néocomien de St. Sauveur-en-Puisaye. Annales de la Societé Entomologique de France, (2) **7**: 95-141, pls. 4, 5.

ROEMER, F. A., 1836. Die Versteinerungen des norddeutschen Oolithen-Gebirges: 1-218, 16 pls. (Im Verlage der Hahn'schen Hofbuchhandlung, Hannover).

— —, 1841. Die Versteinerungen des norddeutschen Kreidegebirges. (Hannover).

——, 1887. *Graptocarcinus texanus*, ein Brachyure aus der Kreide von Texas. Neues Jahrbuch für Mineralogie, Geologie, und Paläontologie, **1887**: 173-176.

ROGER, J., 1944. *Eryoneicus* ? *sahel almae* n. sp. Crustacé Décapode du Sénonien du Liban. Bulletin du Muséum National d'Histoire Naturelle, Paris, (2) **16**: 191-194.

——, 1946. Les invertébrés des couches à poissons du Crétacé supérieur du Liban. Mémoires de la Societé Géologique de France, **51**: 1-92.

ROGER, J. & A. F. DE LAPPARENT, 1945. Une nouvelle espèce de Crustacé Décapode, *Palaeophoberus portlandicus*, découverte dans le Portlandien du Pays de Bray. Bulletin de la Societé Géologique de France, (B) **5** (7-9): 365-374.

ROMAN, F. & P. MAZERAN, 1920. Monographie paléontologique de la faune du Turonien du bassin d'Uchauz et de ses dépendances. Archives du Muséum d'Histoire Naturelle de Lyon, **12**: 1-138, 11 pls.

ROSS, A., J. E. LEWIS & R. J. SCOLARO, 1964. New Eocene decapods from Florida. Quarterly Journal of the Florida Academy of Sciences, **27**: 187-196.

ROSS, A. & R. J. SCOLARO, 1964. A new crab from the Eocene of Florida. Quarterly Journal of the Florida Academy of Sciences, **27**: 97-106.

ROUX, P., 1828-1830. Crustacés de la Méditerranée et de son littoral: i-iv, 1-176, pls. 1-45. (Paris and Marseille).

RUIZ DE GAONA, M., 1943. Nota sobre Crustáceos Decápodos de la cantera del Monte Orobe (Alsasua). Boletin Real Sociedad Española de Historia Natural, **40**: 425-433, pl. 28.

RÜPPELL, E., 1830. Beschreibung und Abbildung von 24 Arten kurzschwänzigen Krabben, also Beitrag zur Naturgeschichte des Rothen Meeres: 1-28, pls. 1-6. (H. L. Brönner, Frankfurt am Main).

RUSCONI, C., 1948. Nuevo plesiosaurio, pez y langustas del mar Jurasico de Mendoza. Revista Museum Historia Natural Mendoza, **11**: 3-12.

SAINT LAURENT, M. DE *see* DE SAINT LAURENT, M.

SAINT MARTIN, J.-P. & P. MÜLLER, 1988. Les Crustacés Décapodes du Messinien d'Oranie (Algérie). Geobios, **21** (2): 251-257, pl. 1.

SAKAI, K., 1982. Revision of Upogebiidae (Decapoda, Thalassinidea) in the Indo-West Pacific Region. Researches on Crustacea, Tokyo, (special number) **1**: 1-106.

——, 1988. A new genus and five new species of Callianassidae (Crustacea: Decapoda: Thalassinidea) from northern Australia. The Beagle, Records of the Northern Territory Museum of Arts and Sciences, **5**: 51-69.

——, 1992. The families Callianideidae and Thalassinidae with the description of two new subfamilies, one new genus and two new species (Decapoda: Thalassinidea). Naturalists, **4**: 1-33.

——, 1999. Synopsis of the family Callianassidae, with keys to subfamilies, genera and species, and the description of new taxa (Crustacea: Decapoda: Thalassinidea). Zoologische Verhandelingen, Leiden, **326**: 1-152.

SAKAI, K., M. TÜRKAY & S.-L. YANG, 2006. Revision of the *Helice/Chasmagnathus* complex (Crustacea: Decapoda: Brachyura). Abhandlungen der Senckenbergischen Naturforschenden Gesellschaft, **565**: 1-76.

SAKAI, T., 1933. A new genus and some new species of crabs from Simoda. Science Reports of Tokyo Bunrika Daig, (B) **1**: 137-144.

——, 1935. New or rare species of Brachyura, collected by the Misago during the zoological survey around the Izu-Peninsula. Science Reports of the Tokyo Bunrika Daigaku, (B) **2** (32): 63-88, figs. 1-17, pls. 6-8.

——, 1937. Studies on the crabs of Japan. II. Oxystomata. Science Reports of the Tokyo Bunrika Daigaku, (B) **3** (Suppl. 2): 67-192, 45 figs., pls. 10-19.

— —, 1963. Description of two new genera and 14 new species of Japanese crabs from the collection of His Majesty the Emperor of Japan. Crustaceana, **5**: 213-233.

— —, 1969. Two new genera and twenty-two new species of crabs from Japan. Proceedings of the Biological Society of Washington, **82**: 243-280.

— —, 1974. Notes from the carcinological fauna of Japan (V). Researches on Crustacea, **6**: 86-102.

— —, 1976. Crabs of Japan and the adjacent seas: 1-773, 251 pls. (Kodansha Ltd., Tokyo).

SAKUMOTO, T., H. KARASAWA & K. TAKAYASU, 1992. Decapod crustaceans from the Miocene Izumo Group, southwest Japan. Bulletin of the Mizunami Fossil Museum, **19**: 441-453.

SALVO, G. DI *see* DI SALVO, G.

SAMOUELLE, G., 1819. The entomologist's useful compendium, or an introduction to the British insects, etc.: 1-496. (T. Boys, London).

SANKARANKUTTY, C., 1962. On Decapoda Brachyura from the Andaman and Nicobar Islands. 2. Family Xanthidae. Journal of the Marine Biological Association of India, **4** (1): 121-150.

SATSANGI, P. P. & U. N. CHANGKAKOTI, 1989. Fossil decapod Crustacea from Upper Sylhet (Prang) Limestone of Jaintia Hills, Meghalaya. Journal of the Geological Society of India, **34**: 277-281.

SATSANGI, P. P. & K. PARIDA, 1980. A new fossil crab from Garo Hills, Meghalaya. Journal of the Geological Society of India, **21**: 302-305.

SAUSSURE, H. DE *see* DE SAUSSURE, H.

SAUVAGE, H. E., 1891. Note sur les Crustacés des terrains jurassiques supérieurs du Boulonnais. Annales des Sciences Naturelles, (Zoologie) (7) **13**: 83-96.

SAVIGNY, J. C., 1816. Mémoires sur les animaux sans vertébrés, **1**. Description et classification des animaux invertébrés et articulés: 1-118. (G. Dufour, Paris).

SAY, T., 1817-1818. An account of the Crustacea of the United States. Journal of the Academy of Natural Sciences of Philadelphia, **1** (1) [1817]: 57-63, 65-80, 97-101, 155-169; (2) [1818]: 235-253, 313-319, 374-401, 423-444, 445-458, pl. 4.

SCHAFHÄUTL, K. E., 1851. Geognostische Untersuchungen des süd-bayerischen Alpengebirges. Geognostische Untersuchungen des Bayerischen Lande, I. Beitrag.

— —, 1863. Südbayerns Lethaea geognostica. Der Kressenberg und die südlich von ihm gelegenen Hochalpen geognostisch betrachtet in ihren Petrefacten. (Leipzig).

SCHAUROTH, C., 1865. Verzeichnis der Versteinerungen im Herzogl. Naturalienkabinett zu Coburg.

SCHLOTHEIM, E. F., 1820. Die Petrefactenkunde. (Gotha).

— —, 1822. Nachträge zur Petrefactenkunde. Beiträge zur näheren Bestimmung der versteinerten und fossilen Krebsarten. (Gotha).

SCHLÜTER, C., 1862. Die Macruren Decapoden der Senon- und Cenomanbildungen Westphalens. Zeitschrift der Deutschen Geologischen Gesellschaft, **14**: 702-749.

— —, 1874. Ueber einige jurassische Crustaceen-Typen in der Oberen Kreide. Verhandlungen des Naturhistorischen Vereins der Preussischen Rheinlande und Westfalens, **31**: 41-55.

— —, 1879. Neue und weniger gekannte Kreide- und Tertiärkrebse des nördlichen Deutschlands. Zeitschrift der Deutschen Geologischen Gesellschaft (Berlin), **31**: 586-615, pls. 13-18.

SCHLÜTER, C. & K. VON DER MARCK, 1868. Neue Fische und Krebse aus der Kreide von Westphalen. Palaeontographica, **15**: 269-305, pls. 51-54.

SCHMITT, W. L., 1924. The macruran, anomuran and stomatopod Crustacea. In: Bijdragen tot de kennis der fauna van Curaçao. Resultaten eener reis van Dr. C. J. van der Horst in 1920. Bijdragen tot de Dierkunde, **23**: 61-81.

— —, 1935. Mud shrimps of the Atlantic coast of North America. Smithsonian Miscellaneous Collections, **93** (2): 1-21, pls. 1-4.

SCHRAM, F. R., 1971. *Litogaster turnbullensis*, a Lower Triassic glypheid decapod Crustacean from Idaho. Journal of Paleontology, **45**: 534-537.

— —, 1986. Crustacea: i-xiv, 1-606. (Oxford University Press, Oxford).

SCHRAM, F. R., R. M. FELDMANN & M. J. COPELAND, 1978. The Late Devonian Palaeopalaemonidae and the earliest decapod crustaceans. Journal of Paleontology, **52**: 1375-1387.

SCHRAM, F. R., S. YAN-BIN, R. VONK & R. S. TAYLOR, 2000. The first fossil stenopodidean. Crustaceana, **73**: 235-242.

SCHUBART, C. D., J. A. CUESTA & D. L. FELDER, 2002. Glyptograpsidae, a new brachyuran family from Central America: larval and adult morphology, and a molecular phylogeny of the Grapsoidea. Journal of Crustacean Biology, **22**: 28-44.

SCHÜTZE, E., 1907. Einige *Glyphea*-Arten aus dem schwäbischen Jura. Jahreshefte des Vereins für Vaterländische Naturkunde in Württemberg, **1907**: 341-354.

SCHWEIGERT, G., 2001. Eine neue Art der Gattung *Antrimpos* Münster (Crustacea, Decapoda, Penaeidae) aus dem Oberjura Süddeutschlands. Stuttgarter Beiträge zur Naturkunde, (B) **307**: 1-33, 6 figs., 8 pls.

— —, 2002. Zwei neue Garnelen (Decapoda: Dendrobranchiata, Eukyphida) aus oberjurassischen Plattenkalken Süddeutschlands. Stuttgarter Beiträge zur Naturkunde, (B) **323**: 1-11, 5 figs.

— —, 2003. *Megachela frickhingeri* n. g. n. sp. (Crustacea: Decapoda: Thalassinidea: Axiidae) aus dem Solnhofener Plattenkalk (Ober-Jura, Bayern). Stuttgarter Beiträge zur Naturkunde, (B) **333**: 1-12, 6 figs.

SCHWEIGERT, G., G. DIETL & M. RÖPER, 2000. Die Panzerkrebse der Familie Erymidae Van Straelen (Crustacea, Decapoda) aus dem Nusplinger Plattenkalk (Ober-Kimmeridgium, Schwäbische Alb) im Vergleich mit fränkischen Vorkommen. Stuttgarter Beiträge zur Naturkunde, (B) **285**: 1-25, 1 fig., 5 pls.

SCHWEIGERT, G., R. M. FELDMANN & M. WULF, 2004. *Macroacaena franconica* n. sp. (Crustacea: Brachyura: Raninidae) from the Turonian of S Germany. Zitteliana, (A) **44**: 61-65, 3 figs.

SCHWEIGERT, G. & A. GARASSINO, 2003. *Malmuncina wulfi* n. g. n. sp., the first uncinid decapod (Crustacea: Decapoda: Astacidea) from the Upper Jurassic. Stuttgarter Beiträge zur Naturkunde, (B) **336**: 1-8, 4 figs.

— — & — —, 2004. New genera and species of shrimps (Crustacea: Decapoda: Dendrobranchiata, Caridea) from the Upper Jurassic lithographic limestones of S Germany. Stuttgarter Beiträge zur Naturkunde, (B) **350**: 1-33, 23 figs.

— — & — —, 2005a. First record of the shrimp genus *Carpopenaeus* Glaessner, 1945 (Crustacea: Decapoda: Carpopenaeidae) from the Jurassic. Neues Jahrbuch für Geologie und Paläontologie, (Monatshefte) **2005**: 490-502, 4 figs.

— — & — —, 2005b. The lobster genus *Squamosoglyphea* Beurlen, 1930 (Crustacea: Decapoda: Glypheidae) in the Upper Jurassic lithographic limestones of southern Germany. Neues Jahrbuch für Geologie und Paläontologie, (Monatshefte) **2005**: 269-288, 5 figs., 1 table.

— — & — —, 2006. News on *Pleopteryx kuempeli* Schweigert and Garassino, an enigmatic shrimp (Crustacea: Decapoda: Caridea: Pleopteryxoidea superfam. nov.) from the Upper Jurassic of S Germany. Neues Jahrbuch für Geologie und Paläontologie, (Monatshefte) **2006**: 449-461, 6 figs.

SCHWEIGERT, G., A. GARASSINO, L. R. HALL, R. B. HAUFF & H. KARASAWA, 2003. The lobster genus *Uncina* Quenstedt, 1851 (Crustacea: Decapoda: Astacidea: Uncinidae) from the Lower Jurassic. Stuttgarter Beiträge zur Naturkunde, (B) **332**: 1-43, 8 figs., 12 pls.

SCHWEIGERT, G., A. GARASSINO & B. RIOU, 2006. First record of *Stenochirus* Oppel, 1861 (Crustacea: Decapoda: Stenochiridae) from the Callovian (Middle Jurassic) of La Voulte-sur-Rhône. Neues Jahrbuch für Geologie und Paläontologie, (Monatshefte) **2006**: 65-77, 7 figs.

SCHWEIGERT, G. & M. RÖPER, 2001. Neue Krebse der Gattung *Palaeastacus* (Crustacea: Decapoda: Erymidae) aus oberjurassischen Plattenkalken Süddeutschlands. Stuttgarter Beiträge zur Naturkunde, (B) **313**: 1-10, 5 figs.

SCHWEITZER, C. E., 2000. Tertiary Xanthoidea (Decapoda: Brachyura) from the Pacific Northwest of North America. Journal of Crustacean Biology, **20** (4): 715-742.

— —, 2001. Additions to the Tertiary decapod fauna of the Pacific Northwest of North America. Journal of Crustacean Biology, **21** (2): 521-537.

— —, 2003. Utility of proxy characters for classification of fossils: an example from the fossil Xanthoidea (Crustacea: Decapoda: Brachyura). Journal of Paleontology, **77**: 1107-1128.

— —, 2005a. The genus *Xanthilites* Bell, 1858 and a new xanthoid family (Crustacea: Decapoda: Brachyura: Xanthoidea): new hypotheses on the origin of the Xanthoidea MacLeay, 1838. Journal of Paleontology, **79**: 277-295.

— —, 2005b. The Trapeziidae and Domeciidae (Decapoda: Brachyura: Xanthoidea) in the fossil record and a new Eocene genus from Baja California Sur, Mexico. Journal of Crustacean Biology, **25** (4): 625-636.

SCHWEITZER, C. E., P. ARTAL, B. VAN BAKEL, J. W. M. JAGT & H. KARASAWA, 2007. Revision of the genus *Titanocarcinus* (Decapoda: Brachyura: Xanthoidea) with two new genera and one new species. Journal of Crustacean Biology, **27**: 278-295.

SCHWEITZER, C. E. & C. B. BOYKO, 2000. First occurrence of the genus *Lophomastix* (Decapoda: Albuneidae) in the fossil record and reconsideration of *Blepharipoda brucei* Rathbun, 1926. Journal of Paleontology, **74** (4): 631-635.

SCHWEITZER, C. E. & R. M. FELDMANN, 1999. Fossil decapod crustaceans of the Pysht Formation (late Oligocene to early Miocene) and Quimper Sandstone (late Eocene), Olympic Peninsula, Washington. Annals of Carnegie Museum, **68**: 215-273.

— — & — —, 2000a. First notice of the Chirostylidae (Decapoda) in the fossil record and new Tertiary Galatheidae (Decapoda) from the Americas. Bulletin of the Mizunami Fossil Museum, **27**: 147-165.

— — & — —, 2000b. New species of calappid crabs from western North America and reconsideration of the Calappidae De Haan sensu lato. Journal of Paleontology, **74**: 230-246.

— — & — —, 2000c. Re-evaluation of the Cancridae Latreille, 1803 (Decapoda: Brachyura) including three new genera and three new species. Contributions to Zoology, **69**: 233-250.

— — & — —, 2000d. New fossil portunids from Washington, U.S.A., and Argentina and a reevaluation of generic and family relationships within the Portunoidea Rafinesque (Decapoda: Brachyura). Journal of Paleontology, **74** (4): 636-653.

— — & — —, 2001a. New Cretaceous and Tertiary decapod crustaceans from western North America. Bulletin of the Mizunami Fossil Museum, **28**: 173-210.

— — & — —, 2001b. Differentiation of fossil Hexapodidae Miers (Decapoda: Brachyura) from similar forms. Journal of Paleontology, **75** (2): 330-345.

— — & — —, 2002. New Eocene decapods (Thalassinidea and Brachyura) from southern California. Journal of Crustacean Biology, **22**: 938-967.

— — & — —, 2008 [imprint 2007]. A new classification for some Jurassic Brachyura (Crustacea: Decapoda: Brachyura: Homolodromioidea): families Goniodromitidae Beurlen, 1932 and Tanidromitidae new family. Senckenbergiana Lethaea, **87**: 119-156.

— — & — —, 2008a. New Eocene hydrocarbon seep decapod crustacean (Anomura: Galatheidae: Shinkaiinae) and its paleobiological implications. Journal of Paleontology, **82**: 1021-1029.

— — & — —, 2008b. Revision of the genus *Laeviprosopon* Glaessner, 1933 (Decapoda: Brachyura: Homolodromioidea: Prosopidae) including two new species. Neues Jahrbuch für Geologie und Paläontologie Abhandlungen, **250** (3): 273-285.

— — & — —, 2009a. Revision of the Prosopinae sensu Glaessner, 1969 (Crustacea: Decapoda: Brachyura) including 4 new families and 4 new genera. Annalen des Naturhistorischen Museums in Wien, (A) **110**: 55-121.

— — & — —, 2009b. New species of Longodromitidae Schweitzer & Feldmann, 2009, from the Ernstbrunn Formation, Late Jurassic (Tithonian), Austria. Annalen des Naturhistorischen Museums in Wien, (A) **111**: 207-224.

— — & — —, 2009c. Revision of the genus *Cyclothyreus* Remeš, 1895 (Decapoda: Brachyura: Dromioidea). Neues Jahrbuch für Geologie und Paläontologie Abhandlungen, **253**: 357-372.

— — & — —, 2009d. Revision of *Gabriella* with new species. Neues Jahrbuch für Geologie und Paläontologie Abhandlungen, **252** (1): 1-16.

SCHWEITZER, C. E., R. M. FELDMANN & C. BONADIO, 2009. A new family of brachyuran (Crustacea: Decapoda: Carpilioidea) from the Eocene of Java, Indonesia. Scripta Geologica, Leiden, **138**: 1-10.

SCHWEITZER, C. E., R. M. FELDMANN, V. ĆOSOVIĆ, R. L. M. ROSS & D. A. WAUGH, 2009. New Cretaceous and Eocene Decapoda (Thalassinidea, Brachyura) from British Columbia, Canada. Annals of Carnegie Museum, **77**: 403-423.

SCHWEITZER, C. E., R. M. FELDMANN, A. ENCINAS & M. SUÁREZ, 2006. New Cretaceous and Eocene Callianassioidea (Thalassinidea, Decapoda) from Algarrobo, Chile. Journal of Crustacean Biology, **26**: 73-81.

SCHWEITZER, C. E., R. M. FELDMANN, J. FAM, W. A. HESSIN, S. W. HETRICK, T. G. NYBORG & R. L. M. ROSS, 2003. Cretaceous and Eocene decapod crustaceans from southern Vancouver Island, British Columbia, Canada: 1-66. (NRC Research Press, Ottawa, Ontario).

SCHWEITZER, C. E., R. M. FELDMANN & P. D. GINGERICH, 2004. New decapods (Crustacea) from the Eocene of Pakistan and a revision of *Lobonotus* A. Milne Edwards, 1864. University of Michigan, Contributions from the Museum of Paleontology, **31** (4): 89-118.

SCHWEITZER, C. E., R. M. FELDMANN, G. GONZALEZ-BARBA & V. ĆOSOVIĆ, 2006. New Decapoda (Anomura, Brachyura) from the Eocene Bateque and Tepetate Formations, Baja California Sur, Mexico. Bulletin of the Mizunami Fossil Museum, **33**: 21-45.

SCHWEITZER, C. E., R. M. FELDMANN, G. GONZÁLES-BARBA & F. J. VEGA, 2002. New crabs from the Eocene and Oligocene of Baja California Sur, Mexico and an assessment of the evolutionary and paleobiogeographic implications of Mexican fossil decapods. Paleontological Society Memoir, **76**: 1-43.

SCHWEITZER, C. E., R. M. FELDMANN & H. KARASAWA, 2007. Revision of the Carcineretidae Beurlen, 1930 (Decapoda: Brachyura: Portunoidea) and remarks on the Portunidae Rafinesque, 1815. Annals of Carnegie Museum, **76** (1): 15-37.
SCHWEITZER, C. E., R. M. FELDMANN & I. LAZĂR, 2007. Decapods from Jurassic (Oxfordian) sponge megafacies of Dobrogea, Romania and reconsideration of *Nodoprosopon* Beurlen, 1928. Neues Jahrbuch für Geologie und Paläontologie, (Abhandlungen) **244**: 99-113.
— —, — — & — —, 2009. Fossil Crustacea (excluding Cirripedia and Ostracoda) in the University of Bucharest Collections, Romania, including new species. Bulletin of the Mizunami Fossil Museum, **35**: 1-14.
SCHWEITZER, C. E., R. M. FELDMANN, A. B. TUCKER & R. E. BERGLUND, 2000. Eocene decapod crustaceans from Pulali Point, Washington. Annals of Carnegie Museum, **69**: 23-67.
SCHWEITZER, C. E., M. ITURRALDE-VINENT, J. L. HETLER & J. VELEZ-JUARBE, 2006. Oligocene and Miocene decapods (Thalassinidea and Brachyura) from the Caribbean. Annals of Carnegie Museum, **75**: 111-136.
SCHWEITZER, C. E. & H. KARASAWA, 2004. Revision of *Amydrocarcinus* and *Palaeograpsus* (Decapoda: Brachyura: Xanthoidea) with definition of three new genera. Paleontological Research, **8** (1): 357-362.
SCHWEITZER, C. E., T. G. NYBORG, R. M. FELDMANN & R. L. M. ROSS, 2004. Homolidae De Haan, 1839 and Homolodromiidae Alcock, 1900 (Crustacea: Decapoda: Brachyura) from the Pacific Northwest of North America and a reassessment of their fossil records. Journal of Paleontology, **78**: 133-149.
SCHWEITZER, C. E. & E. W. SALVA, 2000. First recognition of the Cheiragonidae (Crustacea: Decapoda) in the fossil record and comparison of the family to the Atelecyclidae (Crustacea: Decapoda). Journal of Crustacean Biology, **20** (2): 285-298.
SCHWEITZER, C. E., A. M. SHIRK, V. ĆOSOVIĆ, Y. OKAN, R. M. FELDMANN & I. HOŞGÖR, 2007. New species of *Harpactocarcinus* from the Tethyan Eocene and their paleoecological setting. Journal of Paleontology, **81**: 1091-1100.
SCHWEITZER, C. E., J. VELEZ-JUARBE, M. MARTINEZ, A. C. HULL, R. M. FELDMANN & H. SANTOS, 2008. New Cretaceous and Cenozoic Decapoda (Crustacea: Thalassinidea, Brachyura) from Puerto Rico, United States Territory. Bulletin of the Mizunami Fossil Museum, **34**: 1-15.
SCHWEITZER HOPKINS, C. [E.], E. W. SALVA & R. M. FELDMANN, 1999. Reevaluation of the genus *Xanthosia* Bell, 1863 (Decapoda: Brachyura: Xanthidae) and description of two new species from the Cretaceous of Texas. Journal of Paleontology, **73** (1): 77-90.
SCOTT, R. W., 1970. Paleoecology and paleontology of the Lower Cretaceous Kiowa Formation, Kansas. University of Kansas Paleontological Contributions, **52**: 1-94.
SECRETAN, S., 1961. Une nouvelle espece de Xanthidés au Maroc: *Titanocarcinus meridionalis* nov. sp. Notes de Service Géologique de Maroc, **20** (152): 39-50, pls. 1-3.
— —, 1964. Les Crustacés du Jurassique supérieur et Crétacé de Madagascar. Mémoires du Muséum National d'Histoire Naturelle, Paris, (C) **156**: 1-223.
— —, 1968. *Mecochiria foresti*, nouveau Crustacé Décapode du Jurassique supérieur de Solenhofen. Annales de Paléontologie Invertébrés, **54** (2): 227-238, pls. A, B.
— —, 1970. Crustacés Décapodes du Bas-Congo et de l'enclave de Cabinda. Annales du Musée Royal de l'Afrique Centrale, (IN-8°, Sciences Géologiques) **68**: 77-89, pl. 1.
— —, 1972. Crustacés Décapodes nouveaux du Crétacé supérieur de Bolivie. Bulletin du Muséum National d'Histoire Naturelle, Paris, (3) **49**: 1-15.

——, 1975. Les Crustacés du Monte Bolca. In: Studi e ricerche sui giacimento Terziari di Bolca II. Miscellanea Paleontologica, Museo Civico di Storia Naturale, Verona, **2**: 315-388, pls. 1-37.

——, 1982. *Xanthosia robertsi*, Crustacé Décapode du Crétacé de Madagascar: nouveau nom et nouvelles hypothèses sur son origine. Geobios, **15**: 927-933.

SEGERBERG, K. O., 1900. De Anomura och Brachyura dekapoderna inom Skandinaviens Yngre krita. Geologiska Föreningens i Stockholm Förhandlingar, **22** (5): 347-388, pls. 7-9.

SERÈNE, R., 1964. Redescription du genre *Megaesthesius* Rathbun et dèfinition de Chasmocarcininae, nouvelle sous-famille des Goneplacidae (Decapoda Brachyura). Crustaceana, **7** (3): 175-187.

——, 1965. Guide for curators of brachyuran collections in Southeast Asia: 1-65. (Applied Scientific Research Corporation of Thailand, Bangkok).

——, 1968. The Brachyura of the Indo-west Pacific region. Singapore National Academy of Science, Special Publication, **1**: 33-112, 119-120.

——, 1984. Crustacés Décapodes Brachyoures de l'Océan Indien occidental et de la Mer Rouge. Xanthoidea: Xanthidae et Trapeziidae. Editions de l'ORSTOM, Institut Français de Recherche Scientifique pour le Développement en Coopération, Collection Faune Tropicale, **24**: 1-349, 48 pls. (Paris).

SERÈNE, R. & K. ROMIMOHTARTO, 1969. Observations on the species of *Dorippe* from the Indo-Malayan region. Penelitian Laiit di Indonesia – Marine Research in Indonesia, **9**: 1-35.

SERÈNE, R. & C. L. SOH, 1970. New Indo-Pacific genera allied to *Sesarma* Say, 1817 (Brachyura, Decapoda, Crustacea). Treubia, **27** (4): 387-416, pls. 1-8.

—— & ——, 1976. Brachyura collected during the Thai-Danish Expedition (1966). Phuket Marine Research Center, Research Bulletin, **12**: 1-37.

SERÈNE, R. & A. F. UMALI, 1972. The family Raninidae and other new and rare species of brachyuran decapods from the Philippines and adjacent regions. The Philippine Journal of Science, **99** (1-2): 21-105, pls. 1-9.

SISMONDA, E., 1846. Descrizione dei Pesci e dei Crostacei fossili nel Piemonte. Memorie della Reale Accademia di Scienze di Torino, (2) **10**: 1-88, pls. 1-3.

——, 1861. Appendice alla descrizione dei Pesci e dei Crostacei fossili nel Piemonte. Memorie della Reale Accademia di Scienze di Torino, (2) **19**: 1-24.

SKORIKOV, A. S., 1907. K sistematike evropeiskoaziatskikh Potamobiidae. Ezhegodnik Zoologischeskogo Muzeya Imperatorskoi Akademii Nauk, **12**: 115-118.

SMALLEY, A., 1964. The river crabs of Costa Rica, and the subfamilies of the Pseudothelphusidae. Tulane Studies in Zoology, **12**: 5-13.

SMIRNOV, W., 1929. Decapoda der Fischschichten am Schwarzen Flusse von der Stadt Wladikawkas. Arbeiten der Nord-Kaukasischen Assoziation Wissenschaftlicher Institute, **59**: 1-48, pls. 1-13.

SMITH, S. I., 1869a. In: A. E. VERRILL, On the parasitic habits of Crustacea. American Naturalist, **3**: 239-250.

——, 1869b. Notice of the Crustacea collected by Prof. C. F. Hartt on the coast of Brazil in 1867, list of the described species of Brazilian Podophthalmia. Transactions of the Connecticut Academy of Arts and Sciences, **2**: 1-41.

——, 1870. Ocypodoidea: Notes on American Crustacea, no. 1. Transactions of the Connecticut Academy of Arts and Sciences, **2**: 113-176.

——, 1882. XVII. Report on the Crustacea. Part I. Decapoda. Reports on the dredging, under the supervision of Alexander Agassiz, on the east coast of the United States, during the summer of 1880, by the U.S. Coast Survey Steamer 'Blake', commander J. R. Bartlett,

U.S.N., commanding. Bulletin of the Museum of Comparative Zoology, Harvard College, **10**: 1-108.

— —, 1883. Preliminary report on the Brachyura and Anomura dredged in deep water off the south coast of New England by the United States Fish Commission in 1880, 1881, and 1882. Proceedings of the United States National Museum, **6**: 1-57, pls. 1-6.

— —, 1884. XV. Report on the decapod Crustacea of the Albatross dredgings off the east coast of the United States in 1883. Report of the United States Fish Commission, **10**: 345-426.

SOLLAUD, E., 1913. Nouvelles observations sur les crevettes du genre *Campylonotus* Bate (=*Anchistiella* A.M.E.) type d'une nouvelle famille de *Caridea*: les Campylonotidae. Bulletin du Muséum d'Histoire Naturelle, Paris, **19**: 184-190.

SORGENFREI, T., 1940. Marines Untermiocän im Klintinghoved auf der Insel Als. Ein Beitrag zur Lösung der Aquitanien-Frage. Danmarks Geologiske Undersøgelse, **65**: 90-129, pls. 4-8.

SOWERBY, J., 1826. Description of a new species of *Astacus*, found in a fossil state at Lyme Regis. Zoological Journal, **2**: 493-494.

SPANIER, E. & K. L. LAVALLI, 2006. *Scyllarides* species. In: B. F. PHILLIPS (ed.), Lobsters: biology, management, aquaculture and fisheries: 462-496. (Blackwell Publishing, Oxford, U.K.).

SQUIRES, R. L., 1980. A new species of brachyuran from the Paleocene of California. Journal of Paleontology, **54**: 472-476.

— —, 2001. Additions to the Eocene megafossil fauna of the Llajas Formation, Simi Valley, southern California. Natural History Museum of Los Angeles County Contributions in Science, **489**: 1-40.

SQUIRES, R. L. & R. A. DEMETRION, 1992. Paleontology of the Eocene Bateque Formation, Baja California Sur, Mexico. Natural History Museum of Los Angeles County Contributions in Science, **434**: 1-55.

STAINIER, X., 1887. *Coeloma rupeliense*, Brachyure nouveau de l'argile rupélienne. Annales de la Societé Géologique de Belgique, **14**: 86-96, pl. 5.

STEBBING, T. R. R., 1920. South African Crustacea (Part X of S.A. Crustacea, for the Marine Investigations in South Africa). Annals of the South African Museum, **17** (4): 231-272.

STENZEL, H. B., 1934. Decapod crustaceans from the Middle Eocene of Texas. Journal of Paleontology, **8**: 38-56, pls. 6-7.

— —, 1935. Middle Eocene and Oligocene decapod crustaceans from Texas, Louisiana, and Mississippi. The American Midland Naturalist, **16**: 379-400.

— —, 1944a. A new Cretaceous crab, *Graptocarcinus muiri*, from Mexico. Journal of Paleontology, **18**: 550-552.

— —, 1944b. A new Paleocene catometope crab from Texas, *Tehuacana tehuacana*. Journal of Paleontology, **18**: 546-549.

— —, 1945. Decapod crustaceans from the Cretaceous of Texas. The University of Texas Publication, **4401**: 401-477.

— —, 1952. Decapod crustaceans from the Woodbine Formation of Texas. In: L. W. STEPHENSON (ed.), Larger invertebrate fossils of the Woodbine Formation (Cenomanian) of Texas. Geological Survey Professional Paper, **242**: 212-217, pl. 59.

STEPHENSON, W. & B. CAMPBELL, 1960. The Australian portunids (Crustacea: Portunida) IV. Remaining genera. Australian Journal of Marine and Freshwater Research, **11** (1): 73-122, pls. 1-6.

ŠTEVČIĆ, Z., 2005. The reclassification of brachyuran crabs (Crustacea: Decapoda: Brachyura). Natura Croatica, **14** (Suppl.) (1): 1-159.

STIMPSON, W. [S.], 1857. The Crustacea and Echinodermata of the Pacific shores of North America. Journal of the Boston Society of Natural History, **6**: 1-92, 6 pls.

— —, 1857 [1856-1859] [imprint 1859]. Notices of new species of Crustacea of western North America. Proceedings of the Boston Society of Natural History, **6**: 4-89.

— —, 1858a. Prodromus descriptionis animalium evertebratorum, quae in Expeditione ad Oceanum Pacificum Septentrionalem, a Republica Federata missa, Cadwaladaro Ringgold et Johanne Rodgers ducibus, observavit et descripsit. Pars VII. Crustacea Anomura. Proceedings of the Academy of Natural Sciences of Philadelphia, **10**: 225-252. [Pages 63-90 on separate.]

— —, 1858b. Prodromus descriptionis animalium evertebratorum, quae in Expeditione ad Oceanum Pacificum Septentrionalem, a Republica Federata missa, Cadwaladaro Ringgold et Johanne Rodgers ducibus, observavit et descripsit W. Stimpson. Pars VI. Crustacea Oxystomata. Proceedings of the Academy of Natural Sciences, Philadelphia, **10**: 159-163.

— —, 1858c. Prodromus descriptionis animalium evertebratorum, quae in Expeditione ad Oceanum Pacificum Septentrionalem, a Republica Federata missa, Cadwaladaro Ringgold et Johanne Rodgers ducibus, observavit et descripsit W. Stimpson. Pars V. Crustacea Ocypodoidea. Proceedings of the Academy of Natural Sciences, Philadelphia, **10**: 93-110.

— —, 1859 [read 1858]. Prodromus descriptionis animalium evertebratorum, quae in Expeditione ad Oceanum Pacificum Septentrionalem, a Republica Federata missa, Cadwaladaro Ringgold et Johanne Rodgers Ducibus, observavit et descripsit. Pars IV. Crustacea Cancroidea et Corystoidea. Proceedings of the Academy of Natural Sciences of Philadelphia, **10**: 31-41.

— —, 1860. 6. Sketch of a revision of the genera of Mithracidæ. American Journal of Science and Arts, **29** (85): 132-133.

— —, 1862a. Notes on North American Crustacea, No. 1. Annals of the Lyceum of Natural History of New York, **7**: 49-93, pl. 1 [March 1859]. [Pages 3-47 on separate.]

— —, 1862b. Notes on North American Crustacea, in the Museum of the Smithsonian Institution. No. II. Annals of the Lyceum of Natural History of New York, **7** [1862]: 176-246, pls. 2, 5 [read April 1860]. [Pages 49-118, pls. 2, 3 on separate.]

— —, 1863. On the fossil crab of Gay Head. Boston Journal of Natural History, **7** [1859-63]: 583-589, pl. 12.

— —, 1866. Descriptions of new genera and species of macrurous Crustacea from the coasts of North America. Proceedings of the Chicago Academy of Sciences, **1**: 46-48.

— —, 1870-1881. Brachyura, Pt. 1. In: L. F. DE POURTALÈS (ed.), Preliminary report on the Crustacea dredged in the Gulf Stream in the Straits of Florida. Bulletin of the Museum of Comparative Zoology at Harvard College, **2**: 150-156.

— —, 1871a. Preliminary report on the Crustacea dredged in the Gulf Stream in the Straits of Florida, by L. F. de Pourtalès, Assist. U.S. Coast Survey. Part I. Brachyura. Bulletin of the Museum of Comparative Zoology, **2** (1-5): 109-160.

— —, 1871b. Notes on North American Crustacea, in the Museum of the Smithsonian Institution. No. III. Annals of the Lyceum of Natural History of New York, **10** (4-5): 92-136.

— —, 1873. On some California Crustacea. Proceedings of the California Academy of Natural Science (2nd ed.), **1** [1854-1857]: 1-88.

— —, 1907. Report on the Crustacea (Brachyura and Anomura) collected by the North Pacific Exploring Expedition, 1853-1856. Smithsonian Miscellaneous Collections, **49**: 1-240, 26 pls.

STOLICZKA, F., 1871. Observations of fossil crabs from Tertiary deposits in Sind and Kutch. Memoirs of the Geological Survey of India, Palaeontologica Indica, (7) (14) **1** (1): 1-16, 5 pls.

STOLLEY, E., 1890. Über zwei neue Brachyuren aus dem mitteloligozänen Septarienton Norddeutschlands. Mitteilungen aus dem Mineralogischen Institut der Universität Kiel, **1**: 151-173, pls. 5, 6.

— —, 1924. Über einige Krebsreste aus schweizerischem und norddeutschem Neokom. Eclogae Geologicae Helvetiae, **18**: 411-429, pl. 13.

STRAELEN, V. VAN *see* VAN STRAELEN, V.

STRAND, E., 1928. Miscellanea nomenclatoria zoologica et palaeontologica. I-II. Archiv für Naturgeschichte, **92** (A) (8): 40-41.

STUBBLEFIELD, C. J., 1927. Lower Miocene Crustacea from Pemba Island. Reports on the Paleontology of Zanzibar: 118-120, pls. 22, 23.

— —, 1946. Some decapodan Crustacea from the Middle Eocene of British Somaliland. Annals and Magazine of Natural History, (11) **13**: 505-519, pls. 8-9.

STUDER, T., 1892. Über zwei fossile dekapode Krebse aus den Molasseablagerungen des Belpberges. Abhandlungen der Schweizer Paläontologischen Gesellschaft, **19**.

— —, 1898. Zwei neue Brachyuren aus der miocaenen Molasse. Abhandlungen der Schweizerischen Paläontologischen Gesellschaft, **25**: 1-9, table 1.

SVEN, C., R. H. B. FRAAIJE & G. J. VAN DER ZWAAN, 2001. Polymorphy and extinction of the Late Cretaceous burrowing shrimp *Protocallianassa faujasi* and first record of the genera *Corallianassa* and *Calliax* (Crustacea, Decapoda, Thalassinoidea) from the Cretaceous. Contributions to Zoology, **70**: 85-98.

SZOMBATHY, K., 1916. A *Potamon* (*Telphusa*) nem harmadkori alakjai és palaearktikus utódjaik. Die Tertiären formen der Gattung *Potamon* (*Telphusa*) und ihre paläarktische Nachkommen. Annales Musei Nationalis Hungarici, **14**: 381-421, pl. 3.

TAKEDA, M. & I. FUJIYAMA, 1983. Three decapod crustaceans from the Lower Cretaceous Miyako Group, northern Japan. Bulletin of the National Science Museum, Tokyo, (C, Geology and Paleontology) **9** (4): 129-136, pls. 1, 2.

— — & — —, 1984. A new majid crab from the Miocene Matuyama Group, Saitama Prefecture, central Japan. Bulletin of the National Science Museum, Tokyo, (C, Geology and Paleontology), **10** (2): 49-53.

TAKEDA, M. & S. MIYAKE, 1972. New crabs from the sea around the Tsushima Islands. Bulletin of the National Science Museum, Tokyo, **15** (2): 253-265, figs. 1-5.

TAKEDA, M. & S. TOMIDA, 1984. Two new fossil crabs of the Tymolidae from the Miocene Mizunami Group, central Japan. Bulletin of the Mizunami Fossil Museum, **11**: 39-48, pl. 13.

TAN, S. H. & P. K. L. NG, 2007. Descriptions of new genera from the subfamily Parthenopinae (Crustacea: Decapoda: Brachyura: Parthenopidae). Raffles Bulletin of Zoology, (Suppl.) **16**: 95-119.

TATE, R. & J. C. BLAKE, 1876. The Yorkshire Lias. (London).

TAVARES, M., 1992. Tendences évolutives chez les Crabes primitifs, avec la description d'un nouveau type de chambre incubatrice (Crustacea, Decapoda: Cyclodorippidae Ortmann, 1892, et Xeinostominae subfam. nov.). Comptes Rendus Academie des Sciences, Paris, (3) **314**: 509-514.

— —, 1998. Phyllotymolinidae, nouvelle famille de Brachyoures Podotremata (Crustacea, Decapoda). Zoosystema, **20** (1): 109-122.

TAYLOR, B. J., 1979. Macrurous Decapoda from the Lower Cretaceous of south-eastern Alexander Island. British Antarctic Survey, Scientific Reports, **81**: 1-39.

TAYLOR, R. S., F. R. SCHRAM & Y.-B. SHEN, 1999. A new crayfish family (Decapoda: Astacida) from the Upper Jurassic of China, with a reinterpretation of other Chinese crayfish taxa. Paleontological Research, **3**: 121-136.

TERUZZI, G., 1990. The genus *Coleia* Broderip, 1835 (Crustacea, Decapoda) in the Sinemurian of Osteno in Lombardy. Atti della Società italiana di Scienze naturali e del Museo civico di Storia naturale in Milano, **131** (4): 85-104.

TESSIER, G., C. BESCHIN, A. BUSULINI & A. DE ANGELI, 1999. Nuovi Brachiuri Eocenici nella cava "Main" di Arzignano (Vicenza – Italia settentrionale). Lavori – Società Veneziana Scienze Naturali, **24**: 93-105.

THÉOBALD, N., 1953. Une nouvelle espèce de Crustacé Décapode fossile de l'Aalénien du Bas-Rhin. Comptes Rendus, Academie des Sciences, Societé Géologique de France, **10**: 150-152.

THURMANN, J., 1853. Lettres écrites du Jura à la Société d'Histoire Naturelle de Berne. Lettre XI. Mitteilungen der Naturforschenden Gesellschaft Bern, **265**: 1-13.

TIWARI, K. K., 1963. Lower Tertiary penaeid shrimps from Kapurdi (Barmer District, Rajasthan, India). Crustaceana, **5**: 205-212.

TODD, J. A. & J. S. H. COLLINS, 2005. Neogene and Quaternary crabs (Crustacea, Decapoda) collected from Costa Rica and Panama by members of the Panama Paleontology Project. Bulletin of the Mizunami Fossil Museum, **32**: 53-85, pls. 1-4.

TORABY, H. & M. YAZDI, 2002. First report on Miocene decapod fauna (Crustacea) from Central Iran, a prelimenary [sic] study on their environmental and ecological factors. Geological Society of Australia, Abstract, **68**: 156.

TOULA, F., 1904. Über eine neue Krabbe (*Cancer Bittneri* n. sp.) aus dem miocänen Sandsteine von Kalksburg bei Wien. Jahrbuch der Kaiserlich-Königlichen Geologischen Reichsanstalt, **54**: 161-168.

TRAUTSCHOLD, H., 1866. Zur Fauna des russischen Jura. Bulletin Société Impériale des Naturalistes de Moscou, **39**: 1-24, pls. 1-4.

TRIBOLET, M. DE *see* DE TRIBOLET, M.

TSHUDY, D. & L. E. BABCOCK, 1997. Morphology-based phylogenetic analysis of the clawed lobsters (family Nephropidae and the new family Chilenophoberidae). Journal of Crustacean Biology, **17**: 253-263.

TSHUDY, D., W. S. DONALDSON, C. COLLOM, R. M. FELDMANN & C. E. SCHWEITZER, 2005. *Hoploparia albertaensis*, a new species of clawed lobster (Nephropidae) from the late Coniacian, shallow-marine Bad Heart Formation of northwestern Alberta, Canada. Journal of Paleontology, **79** (5): 961-968.

TSHUDY, D. M. & U. SORHANNUS, 2000. *Jagtia kunradensis*, a new genus and species of clawed lobster (Decapoda: Nephropidae) from the Upper Cretaceous (Upper Maastrichtian) Maastricht Formation, the Netherlands. Journal of Paleontology, **74** (2): 224-229.

TUCKER, A. B., 1998. Systematics of the Raninidae (Crustacea: Decapoda: Brachyura), with accounts of three new genera and two new species. Proceedings of the Biological Society of Washington, **111**: 320-371.

TUCKER, A. B. & R. M. FELDMANN, 1990. Fossil decapod crustaceans from the Lower Tertiary of the Prince William Sound region, Gulf of Alaska. Journal of Paleontology, **64** (3): 409-427.

TUCKER, A. B., R. M. FELDMANN & C. L. POWELL, II, 1994. *Speocarcinus berglundi* n. sp. (Decapoda: Brachyura), a new crab from the Imperial Formation (late Miocene-late Pliocene) of southern California. Journal of Paleontology, **68** (4): 800-807.

VAN BAKEL, B. W. M., P. ARTAL, R. H. B. FRAAIJE & J. W. M. JAGT, 2009. A new Early Oligocene crab (Decapoda, Brachyura, Dromiacea) from northwest Belgium, with comments on its palaeobiology. Geologica Belgica, **12**: 45-57.

VAN BAKEL, B. W. M., R. H. B. FRAAIJE & J. W. M. JAGT, 2005. A new Early Campanian species of *Xanthosia* (Decapoda, Brachyura) from northwestern Germany. Crustaceana, **78** (3): 285-295, figs. 1-3.

VAN BAKEL, B. W. M., R. H. B. FRAAIJE, J. W. M. JAGT & P. ARTAL, 2008. An unexpected diversity of Late Jurassic hermit crabs (Crustacea, Decapoda, Anomura) in Central Europe. Neues Jahrbuch für Geologie und Paläontologie, Abhandlungen, **250**: 137-156.

VAN BAKEL, B. W. M., J. W. M. JAGT, P. ARTAL & R. H. B. FRAAIJE, 2009. *Harenacorystes johanjansseni*, a new Pliocene crab (Crustacea, Decapoda) from the Netherlands, and notes on Miocene-Pliocene corystoid crabs from the North Sea Basin. Bulletin of the Mizunami Fossil Museum, **35**: 79-85.

VAN BAKEL, B. W. M., J. W. M. JAGT & R. H. B. FRAAIJE, 2002. *Ciliopagurus obesus*, a new species of Oligocene hermit crab from northwest Belgium. Cainozoic Research, **2** (1-2): 109-116, 2 figs.

— —, — — & — —, 2003. A new Pliocene cancrid crab from Oelegem, province of Antwerpen (NW Belgium). Cainozoic Research, **2** (1-2) [2002]: 79-85.

VAN BAKEL, B. W. M., J. W. M. JAGT, R. H. B. FRAAIJE & E. R. H. WILLE, 2003. Piacenzian (Pliocene) decapod crustacean faunules from northwest Belgium. Bulletin of the Mizunami Fossil Museum, **30**: 97-108, 6 pls.

VAN BENEDEN, P. J., 1872. Sur la découverte d'un homard fossile dans l'argile de Rupelmonde. Bulletin de l'Academie Royale de Belgique, (2) **33**: 316-321.

VAN BINKHORST, J. T., 1857. Neue Krebse aus der Maestrichter Tuffkreide. Verhandlungen des Naturhistorischen Vereins der Preussischen Rheinlande und Westfalens, **14**: 107-110, pls. 6, 7.

VAN HOEPEN, E. C. N., 1962. An Albian astacurid from Zululand. Annale van die Geologiese Opname, Pretoria, **1**: 253-255.

VAN STRAELEN, V., 1921a. Sur des Crustacés du Jurassique de la Nièvre. Bulletin de la Société Belge de Géologie, de Paléontologie et d'Hydrologic, **20**: 139-142, pl. 1.

— —, 1921b. *Hoploparia corneti* Crustacé Décapode nouveau de l'Yprésian supérieur de Cuesmes. Bulletin de la Société Belge de Géologie, Paléontologie et d'Hydrologie, **30**: 136-138.

— —, 1922a. Les Crustacés Décapodes du Callovien de la Voulte-sur-Rhône (Ardèche). Comptes Rendus Académie des Science, Paris, **175**: 982-983.

— —, 1923a. Description de Crustacés Décapodes Macroures nouveaux des terrains secondaires. Annales de la Société Royale Zoologique de Belgique, **53** [1922]: 84-93.

— —, 1923b. Sur les Crustacés Décapodes du Bathonien. Comptes Rendus Académie des Sciences, Paris, **177**: 552-554.

— —, 1924 [imprint 1923]. Description des Raniniens nouveaux des terrains tertiairs de Bornéo. Proceedings Koninklijke Nederlandse Akademie van Wetenschappen, Amsterdam, **26**: 777-782.

— —, 1924. *Portunus brouweri*, Portunien nouveau de tertiaire de l'Ile Célèbes. Jaarboek van het Mijnwezen in Ned.-Oost Indië, Verhandelingen, **1924**: 168-171, fig. 1.

— —, 1924 [imprint 1925]. Contribution à l'étude des Crustacés Décapodes de la période jurassique. Mémoires d'Académie Royale de Belgique, (Science), [collected in number 4, series 2 = (2) **4**] **7**: 1-462, pls. 1-10.

——, 1925. Description de Brachyoures montiens du Cominges. Bulletin de la Société Belge de Géologie, **34**: 58-62.
——, 1927. Contribution à l'étude des Crustacés Décapodes fossiles de la Péninsule Ibérique. Revista Española de Entomología, **3** (1): 79-94, pls. 1-4.
——, 1928a. Sur les Crustacés Décapodes triasiques et sur l'origine d'un phylum de Brachyoures. Bulletin du Musée Royal d'Histoire Naturelle de Belgique, Bruxelles, **14**: 496-516.
——, 1928b. *Astacus Edwardsi* Munier-Chalmas Ms., Stacidae du Paléocène de Sézanne (Champagne). Bulletin de la Société Géologique de France, (4) **28**: 3-6.
——, 1928c. On a fossil freshwater crayfish from eastern Mongolia. Bulletin of the Geological Society of China, Peking, **7**: 133-138.
——, 1929. Sur des Crustacés Décapodes jurassiques du Groenland oriental. Bulletin de l'Académie Royale de Belge, (Science), Bruxelles, (5) **15** (10): 741-745.
——, 1929 [1930]. *Xanthopsis Cuvillieri*, Xanthide nouveau de l'Éocène supérieur de l'Égypte. Bulletin de la Société Belge de Géologie, de Paléontologie et d'Hydrologie, **39**: 157-158.
——, 1933a. *Antrimpos madagascariensis* Crustacé Décapode du Permotrias de Madagascar. Bulletin du Musée Royal d'Histoire Naturelle de Belgique, **9** (15): 1-3.
——, 1933b. Sur des Crustacés Décapodes Cénozoïques de Venezuela. Bulletin du Musée Royal d'Histoire Naturelle de Belgique, **9** (10): 1-11.
——, 1933c. Sur des Crustacés Décapodes de l'Eocène supérieur de l'Ile Bonaire. Bulletin du Musée Royal d'Histoire Naturelle de Belgique, **9** (23): 1-4.
——, 1934. Contribution à l'étude des Crustacés Décapodes fossiles de la Catalogne. Géologia Pays Catalans, **3** (25): 1-6.
——, 1935 [imprint 1934]. Sur un Raninide nouveau du Danien de la Libye. Bulletin de l'Institut d'Egypte, **17**: 113-115.
——, 1936a. Crustacés Décapodes nouveaux ou peu connus de l'époque Crétacique. Bulletin du Musée Royal d'Histoire Naturelle de Belgique, **12** (45): 1-49.
——, 1936b. Sur les Crustacés Décapodes triasiques du Nevada. Bulletin du Musée Royal d'Histoire Naturelle de Belgique, Bruxelles, **12** (29): 1-7.
——, 1936c. Sur des Crustacés Décapodes de l'Auversien des Alpes de Schwyz. Bulletin du Musée Royal d'Histoire Naturelle de Belgique, Bruxelles, **12** (27): 1-6.
——, 1936d. Sur quelques Crabes du Sahélien moyen des environs d'Oran. Bulletin de la Société Géologique de France, (5) **6** (6-8): 473-480, pl. 33.
——, 1937. *Parapirimela angolensis*, Brachyure nouveau du Miocène de l'Angola. Bulletin du Musée Royal d'Histoire Naturelle de Belgique, Bruxelles, **8** (5): 1-4.
——, 1938. Crustacés Décapodes cénozoiques des Indes Orientales Néerlandaises. Leidsche Geologische Mededeelingen, **10** (1): 90-103.
——, 1939a. *Martinezicancer schencki*, Brachyure nouveau du Paléocène de Californie. Bulletin du Musée Royal d'Histoire Naturelle de Belgique, Bruxelles, **15** (63): 1-3.
——, 1939b. *Neptunus regulensis* nov. sp. Portunidae nouveau du Rupélien de l'Aquitaine. Bulletin du Musée Royal d'Histoire Naturelle de Belgique, **15** (58): 1-3.
——, 1940a. Pénéides nouveaux Méso- et Cénozoiques. Bulletin du Musée Royal d'Histoire Naturelle de Belgique, **16** (8): 1-5.
——, 1940b. Crustacés Décapodes nouveaux du Crétacique de la Navarre. Bulletin du Musée Royal d'Histoire Naturelle de Belgique, **16** (4): 1-5, pl. 1.
——, 1944. Anomure et Brachyures du Cénomanien de la Navarre. Bulletin du Musée Royal d'Histoire Naturelle de Belgique, **20** (25): 1-12, pl. 1.

— —, 1949. Crustacés. In: M. COLLIGNON (ed.), Recherches sur les faunes albiennes de Madagascar, I, L'Albien d'Ambarimaninga. Annales Géologiques de la Service des Mines Madagascar, **16**: 1-99.

VEGA, F. J. & M. DEL CARMEN PERRILLIAT, 1989. Una especie nueva de cancrejo del género *Costacopluma* (Crustacea: Decapoda: Retroplumidae) del Maastrichtiano de Nuevo León, México. Universidad Nacional Autónoma, Instituto de Geología, Revista, **8**: 84-87.

VEGA, F. J., T. COSMA, M. A. COUTIÑO, R. M. FELDMANN, T. G. NYBORG, C. E. SCHWEITZER & D. A. WAUGH, 2001. New Middle Eocene decapods (Crustacea) from Chiapas, Mexico. Journal of Paleontology, **75**: 929-946.

VEGA, F. J. & R. M. FELDMANN, 1991. Fossil crabs (Crustacea: Decapoda) from the Maastrichtian Difunta Group, northeastern Mexico. Annals of Carnegie Museum, **60** (2): 163-177.

— — & — —, 1992. Occurrence of *Costacopluma* (Decapoda: Brachyura: Retroplumidae) in the Maastrichtian of southern Mexico and its paleobiogeographic implications. Annals of Carnegie Museum, **61** (2): 133-152.

VEGA, F. J., R. M. FELDMANN, P. GARCÍA-BARRERA, H. FILKORN, F. PIMENTEL & J. AVENDAÑO, 2001. Maastrichtian Crustacea (Brachyura: Decapoda) from the Ocozocuautla Formation in Chiapas, southeast Mexico. Journal of Paleontology, **75**: 319-329.

VEGA, F. J., R. M. FELDMANN, A. O. OCAMPO & K. O. POPE, 1997. A new species of Late Cretaceous crab (Brachyura: Carcineretidae) from Albion Island, Belize. Journal of Paleontology, **71** (4): 615-620.

VEGA, F. J., R. M. FELDMANN & F. SOUR-TOVAR, 1995. Fossil crabs (Crustacea: Decapoda) from the Late Cretaceous Cárdenas Formation, east-central Mexico. Journal of Paleontology, **69** (2): 340-350.

VEGA, F. J., R. M. FELDMANN & J. L. VILLALOBOS-HIRIART, 1995. Additions to the crustacean (Decapoda) fauna from the Potrerillos Formation (Late Cretaceous) in northeastern Mexico. Annals of Carnegie Museum, **64** (3): 39-49.

VEGA, F. J., R. M. FELDMANN, J. L. VILLALOBOS-HIRIART & R. GÍO-ARGAEZ, 1999. A new decapod fauna from the Miocene Tuxpan Formation, eastern Mexico. Journal of Paleontology, **73** (3): 407.

VEGA, F. J., P. GARCÍA-BARRERA, M. C. PERRILLIAT, M. A. COUTIÑO & R. MARIÑO-PÉREZ, 2006. El Espinal, a new plattenkalk facies locality from the Lower Cretaceous Sierra Madre Formation, Chiapas, southeastern Mexico. Revista Mexicana de Ciencias Geológicas, **23** (3): 323-333.

VEGA, F. J., T. NYBORG, M. A. COUTIÑO & O. HERNÁNDEZ-MONZÓN, 2008. Review and additions to the Eocene decapod Crustacea from Chiapas, Mexico. Bulletin of the Mizunami Fossil Museum, **34**: 51-71.

VEGA, F. J., T. NYBORG, R. H. B. FRAAYE & B. ESPINOSA, 2007. Paleocene decapod Crustacea from the Rancho Nuevo Formation (Parras Basin-Difunta Group), northeastern Mexico. Journal of Paleontology, **81** (6): 1423-1441.

VEIGA FERREIRA, O. DA, 1954. Malacostraceos do Miocénico marinho de Portugal. Comunicações dos Serviços Geologicos de Portugal, **35**: 5-23, pls. 1-6.

— —, 1961. Nove espécie de *Calianassa* no Miocénico da Bacia do Tejo. Comunicações dos Serviços Geologicos de Portugal, **45**: 479-481, pl. 1.

VERESHCHAKA, A. L., 1997. New family and superfamily for a deep-sea shrimp from the Galathea collections. Journal of Crustacean Biology, **17**: 361-373.

VERHOFF, J., P. MÜLLER, R. M. FELDMANN & C. E. SCHWEITZER, 2009. A new species of Tumidocarcinidae (Decapoda, Carpilioidea) from the Kambühel Formation (Paleocene) of Austria. Annalen des Naturhistorischen Museums in Wien, (A) **111**: 225-232.

VERMA, K. K., 1977. Cancroid crabs from the Quilon Beds (Lower Miocene) of Kerala, India. Journal of the Palaeontological Society of India, **20**: 305-313.

VIA, L., 1932. Els crancs fòssils del Terciari de Catalunya. Butlletí de la Institució Catalana d'Història Natural, **32** (4): 131-146, 2 pls.

— —, 1941. Los cangrejos fósiles de Cataluña. Boletín del Insituto Geológico y Minero de España, **55**: 3-73, pls. 1-11.

— —, 1948. A propósito de dos Braquiuros fósiles del Eoceno de la provincia de Barcelona. Estudios Geológicos, **4** (8): 283-286, 4 pls.

— —, 1951. Contribución al estudio de los decápodos del secundario en España. Anales de la Escuela de Peritos Agrícolas y Superior de Agricultura y de los Servicios Técnicos de Agricultura, **10**: 151-180, pl. 1.

— —, 1959. Decápodos fósiles del Eoceno español. Boletín del Instituto Geológico y Minero de España, **70**: 1-72.

— —, 1966. *Pinnixa* (*Palaeopinnixa*) *mytilicola*, nuevo braquiuro fósil, en el mioceno marino del Vallés (Barcelona). Acta Geológica Hispánica, **1** (4): 1-4.

— —, 1971. Crustáceos decápodos del jurásico superior del Montsec (Lérida). Cuadernos Geología Ibérica, **2**: 607-612.

— —, 1981. Les Crustacés Décapodes du Cénomanien de Navarra (Espagne): premiers resultats de l'étude des Galatheidae. Géobios, **14** (2): 247-251.

VINASSA DE REGNY, II, P., 1897. Contribuzioni alla conoscenza dei Crostacei fossili italiani. *Simoniellia quiricensis* n. g. n. sp. del Pliocene di San Quirico d'Orcia. Rivista Italiana di Paleontologia, **3** (5-6): 19-24, 1 pl.

VOLTZ, F., 1835. *Palinurus Münsteri*. Neues Jahrbuch für Mineralogie, Geognosie, Geologie und Petrefaktenkunde, **1835**: 62.

VON ALBERTI, F., 1864. Überblick über die Trias mit Berücksichtigung ihres Vorkommens in den Alpen. (Stuttgart).

VON ALBRECHT, H., 1983. Die Protoastacidae n. fam., fossile Vorfahren der Flusskrebse? Neues Jahrbuch für Geologie und Paläontologie, Monatshefte, Stuttgart, **1983** (1): 5-15.

VON AMMON, L., 1905. Zur Geologie von Togo und vom Nigerlande. Mitteilungen der Geographischen Gesellschaft in München, **1** (3): 393-474.

VON DER MARCK, W., 1858. Über einige Wirbeltiere, Kruster und Cephalopoden der westfälischen Kreide. Zeitschrift der Deutschen Geologischen Gesellschaft, **10**: 231-271.

— —, 1863. Fossile Fische, Krebse und Pflanzen aus dem Plattenkalke der jüngsten Kreide in Westphalen. Palaeontographica, **11**: 1-83.

VON FISCHER-BENZON, R., 1866. Über das relative Alter des Faxoe-Kalkes und über die in demselben vorkommenden Anomuren und Brachyuren. (Kiel).

VON FRITSCH, K., 1870. Vorstudien über die jüngeren mesozoischen Ablagerungen bei Eisenach. Neues Jahrbuch für Mineralogie, Geologie und Palaentologie, **1870**: 385-416, pl. 4.

— —, 1871. Über einige fossile Crustaceen aus dem Septarienton des Mainzer Beckens. Zeitschrift der Deutschen Geologischen Gesellschaft, **23**: 679-701, pls. 16, 17.

VON GISTL, J. N. F. X., 1848. Naturgeschichte des Thierreichs für Höhere Schulen. (Hoffman'scher Verlags-Buchhandlung).

VON KNEBEL, R., 1907. Die Eryoniden des oberen Weissen Jura von Süddeutschland. Archiv für Biontologie, **2**: 195-233.

VON MARTENS, E., 1872. Über Cubanische Crustaceen nach den Sammlungen Dr. J. Gundlach's. Archiv für Naturgeschichte, **38**: 257-258, pls. 4, 5.

VON MEYER, H., 1834. Krebse im bunten Sandstein. Museum Senckenbergianum, **1**: 293-295.

——, 1835. Briefliche Mitteilungen. Neues Jahrbuch für Mineralogie, Geologie, Geognosie und Petrefaktenkunde, **1834**: 329.
——, 1836. Beiträge zu *Eryon*, einem Geschlechte fossiler langschwänziger Krebse. Nova Acta Physio-Medica Academiae Caes. Leopoldino-Carolinae Naturae Curiosorum, Halle, **18**: 263-283.
——, 1837. Briefliche Mitteilungen. Neues Jahrbuch für Mineralogie, Geologie, Geognosie und Petrefaktenkunde, **1837**: 314.
——, 1840a. Neue Gattungen fossiler Krebse aus Gebilden vom bunten Sandsteine bis in die Kreide: 1-23. (E. Schweizerbart, Stuttgart).
——, 1840b. Briefliche Mitteilungen. Neues Jahrbuch für Mineralogie, Geologie, Geognosie und Petrefaktenkunde, **1840**: 587.
——, 1841. Briefliche Mitteilungen. Neues Jahrbuch für Mineralogie, Geologie, Geognosie und Petrefaktenkunde, **1840**: 96.
——, 1842. Über die in dem dichten Jurakalk von Aalen in Würtemburg vorkommenden Spezies des Crustaceengenus *Prosopon*. Beiträge zur Petrefaktenkunde, **5**: 70-75, pl. 15.
——, 1843. Briefliche Mitteilungen. Neues Jahrbuch für Mineralogie, Geologie, Geognosie und Petrefaktenkunde, **1843**: 589.
——, 1847a. *Halicyne* und *Litogaster*, zwei Crustaceengattungen aus dem Muschelkalke Württembergs. Palaeontographica, **1**: 134-140.
——, 1847b. *Selenisca* und *Eumorphia*, zwei Krebse aus der Oolithgruppe Württemberg's. Palaeontographica, **1**: 141-148.
——, 1847c. *Cancer Paulino-Württembergensis* aus einem jungern Kalkstein in Aegypten. Palaeontographica, **1**: 91-98, pl. 11.
——, 1851a. Fische, Crustaceen, Echinodermen und anderen Versteinerungen aus dem Muschelkalk Oberschlesiens. Palaeontographica, **1**: 216-279.
——, 1851b. Briefliche Mitteilungen. Neues Jahrbuch für Mineralogie, Geologie, Geognosie und Petrefaktenkunde, **1851**: 677.
——, 1856. Briefliche Mitteilungen. Jahrbuch für Mineralogie, Geologie, Geognosie und Petrefaktenkunde, **1856**: 51.
——, 1857. Briefliche Mitteilungen. Jahrbuch für Mineralogie, Geologie, Geognosie und Petrefaktenkunde, **1857**: 556.
——, 1858. Briefliche Mitteilungen. Jahrbuch für Mineralogie, Geologie, Geognosie und Petrefaktenkunde, **1858**: 59-62.
——, 1859. *Eryon raiblanus* aus den Raibler-Schichten in Kärnthen. Palaeontographica, **8**: 27-30.
——, 1860. Die Prosoponiden oder die Familie der Maskenkrebse. Palaeontographica, **7**: 183-222, pl. 23.
——, 1862. Tertiaere Decapoden aus den Alpen, von Oeningen und dem Taunus. Palaeontographica, **10**: 147-178, pls. 16-19.
——, 1864. Briefliche Mitteilungen. Neues Jahrbuch für Mineralogie, Geologie, Geognosie und Petrefaktenkunde, **1864**: 20.
VON MÜCKE, K., 1915. Beitrag zur Kenntnis des Karpathensandsteins im siebenbürgischen Erzgebirge. Verhandlungen der K.-K. Geologischen Reichsanstalt, **8**: 154-162.
VON REUSS, R., 1871. *Phymatocarcinus speciosus*, eine neue fossile Krabbe aus dem Leithakalke des Wiener Beckens. Sitzungsberichte der Akademie in Wien, (Mathematisch-naturwissenschaftliche Klasse) **63-64** (1-5): 325-330, pl. 1.
VON ZITTEL, K. A., 1885. Molluska und Arthropoda. Handbuch der Paläontologie, **2** (2): 679-721. (R. Oldenbourg, Munich).

VOSINSKY, A., 1848. Notice sur les restes des Crustacés fossiles du Jura de Moscou. Bulletin de la Société Impériale des Naturalistes de Moscou, **9**: 494-504.

WALKER, A. O., 1887. Notes on a collection of Crustacea from Singapore. Journal of the Linnean Society of London, **20** (118): 107-117, pls. 6-9.

WANG, Y., 1981. Late Lower Cretaceous fossil Decapoda from Lhasa Region, Xizang. In: The Scientific Expedition to the Qinghai-Xizang Plateau (compilor), 1981. Palaeontology of Xizang, **3**: 349-354. (Science Press, Beijing).

WARD, M., 1936. Crustacea Brachyura from the coasts of Queensland. Memoirs of the Queensland Museum, **11** (1): 1-13, 3 pls.

— —, 1942. Notes on the Crustacea of the Desjardins Museum, Mauritius Institute, with descriptions of new genera and species. Bulletin of the Mauritius Institute, Port Louis, **2** (2): 49-113, pls. 5, 6.

WEBER, F., 1795. Nomenclator entomologicus secundum Entomologiam Systematicum ill. Fabricii adjectis speciebus recens detectis et varietatibus: 1-171. (C. E. Bohn, Chilonii et Hamburgi).

WEHNER, G., 1988. Über die Prosoponiden (Crustacea, Decapoda) des Jura: 1-154, 8 pls., 1 insert. (Dissertation zur Erlangung des Doktorgrades der Fakultät für Geowissenschaften der Ludwig-Maximilians-Universität zu München).

WELLER, S., 1903. The Stokes collection of Antarctic fossils. Journal of Geology, **11**: 413-419.

— —, 1905. The fauna of the Cliffwood (N. J.) Clays. Journal of Geology, **13**: 324-337.

WHETSTONE, K. N. & J. S. H. COLLINS, 1982. Fossil crabs (Crustacea: Decapoda) from the Upper Cretaceous Eutaw Formation of Alabama. Journal of Paleontology, **56**: 1218-1222.

WHITE, A., 1842. Description of an orthopterous Insect, and two new species of Crustacea, from New Zealand in the collection of the British Museum. In: J. E. GRAY (ed.), Zoological Miscellany: 78-79. [1831-1844.]

— —, 1846. Notes on four new genera of Crustacea. Annals and Magazine of Natural History, **18** (118): 176-178, pl. 2, figs. 1-3.

— —, 1847a. List of the specimens of Crustacea in the collection of the British Museum: 1-143. (British Museum, London).

— —, 1847b. Descriptions of new Crustacea from the Eastern Seas. Proceedings of the Zoological Society of London, **15**: 56-58.

— —, 1847c. Descriptions of a new genus and five new species of Crustacean. In: J. B. JUKES (ed.), Narrative of the surveying voyage of H. M. S. Fly, commanded by Captain, F. P. Blackwood, R. N. in Torres Strait, New Guinea, and other islands of the Eastern Archipelago, during the years 1842-1846: together with an excursion into the interior of the eastern part of Java, **2**: 335-338, pl. 2. (T. and W. Boone, London).

— —, 1847d. Short descriptions of some new species of Crustacea in the collection of the British Museum. Proceedings of the Zoological Society of London, **15**: 84-86.

— —, 1852. Descriptions of some new species of Annulosa. In: J. MACGILLIVRAY (ed.), Narrative of the voyage of H. M. S. "Rattlesnake" during the years 1846-1850, **2** (appendix): 387-395. (Boone, London).

— —, 1856. Some remarks on Crustacea of the genus *Lithodes* with a brief description of a species apparently hitherto unrecorded. Proceedings of the Zoological Society of London, **1856**: 132-135, pl. 42.

WHITE, C. A., 1880. Description of new invertebrate fossils from the Mesozoic and Cenozoic rocks of Arkansas, Wyoming, Colorado, and Utah. Proceedings of the United States National Museum, **3**: 157-162.

— —, 1883. Contributions to invertebrate paleontology, no. 2: Cretaceous fossils of the Western States and Territories. In: F. V. HAYDEN, 12th Annual Report of the United States Geological and Geographical Survey of the Territories: a report of progress of the exploration in Wyoming and Idaho for the year 1878. In two parts. Part **1**: 5-173, 42 pls.

WHITEAVES, J. F., 1874. On recent deep-sea dredging operations in the Gulf of St. Lawrence. American Journal of Science, **3** (7): 210-219.

— —, 1885. Report on the Invertebrata of the Laramie and Cretaceous rocks in the vicinity of the Bow and Belly Rivers and adjacent localitites in the North-West Territory. Geological and Natural History Survey of Canada Contribution to Canadian Palaeontology, **1** (1): 87-89, pl. 11.

— —, 1887. On some fossils from the Cretaceous and Laramie Rocks of the Saskatchewan and its tributaries. Geology and Natural History Survey of Canada, Annual Report, (new series) **2** [1887]: 152E-166E.

— —, 1895. On some fossils from the Nanaimo group of Vancouver Cretaceous. Transactions of the Royal Society of Canada, (2) **1** (section 4): 119-133.

WHITFIELD, R. P., 1880. Notice of new forms of fossil crustaceans from the Upper Devonian rocks of Ohio, with descriptions of new genera and species. American Journal of Science, (3) **19**: 33-42.

— —, 1907. Notice of an American species of the genus *Hoploparia* McCoy from the Cretaceous of Montana. Bulletin of the American Museum of Natural History, **23**: 459-462.

WILCKENS, O., 1907. Die Lamellibranchiaten, Gastropoden u.s.w. der Oberen Kreide Süd-patagoniens. Berichte der Naturforschenden Gesellschaft zu Freiburg, Breisgau, **15**: 97-166.

WILLIAMS, A. B., 1980. A new crab family from the vicinity of submarine vents on the Galapagos rift (Crustacea: Decapoda: Brachyura). Proceedings of the Biological Society of Washington, **93** (2): 443-472.

WINKLER, T.-C., 1883. Étude carcinologique sur les genres "*Pemphix*", "*Glyphea*" et "*Araeosternus*". Archives du Musée Teyler, (2) **1**: 73-124.

WINTLE, S. H., 1886. Notes on fossil crabs from the deep dredgings of the Yarra River, and excavation of the Coode Canal, Victoria. Papers and Proceedings of the Royal Society of Tasmania, **1886**: 233-234.

WITHERS, T. H., 1922. On a new brachyurous crustacean from the Upper Cretaceous of Jamaica. Annals and Magazine of Natural History, (9) **10**: 534-541.

— —, 1924a. Some Cretaceous and Tertiary decapod crustaceans from Jamaica. Annals and Magazine of Natural History, (9) **13**: 81-93, pls. 2-4.

— —, 1924b. Some decapod crustaceans (*Callianassa* and *Ranina*) from the Oligocene of Washington State, U.S.A. Annals and Magazine of Natural History, (9) **14**: 121-127, pl. 4.

— —, 1924c. Decapod crustaceans from the Oligocene of Anguilla. Annals and Magazine of Natural History, (9) **14**: 225-233, pl. 6.

— —, 1924d. Eocene brachyurous decapod Crustacea from Nigeria. Annals and Magazine of Natural History, (9) **13**: 94-97, pl. 5.

— —, 1925. *Typilobus sadeki*, a new crab from the Miocene of Sinai. Appendix to F. W. MOON & H. SADEK, Preliminary geologic report on Gebel Khosera area. Ministry of Finances Egypt, Petroleum Research Bulletin, **9**: 37-40, pl. 3.

— —, 1926. Decapod Crustacea (*Callianassa*) from the Scotland beds of Barbados. Geological Magazine, **63** (3): 104-108, pl. 9.

— —, 1927. *Ranina trechmanni*, a new Cretaceous crab from Jamaica. Geological Magazine, **64**: 176-180, pl. 7.

— —, 1928. New Cretaceous crabs from England and Syria. Annals and Magazine of Natural History, (10) **2**: 457-461, pl. 13.

— —, 1932a. A Liassic crab, and the origin of the Brachyura. Annals and Magazine of Natural History, (10) **9**: 313-323, pls. 9, 10.

— —, 1932b. Some Eocene crabs from Persia and India. Annals and Magazine of Natural History, (10) **9**: 467-472.

— —, 1946. New Cretaceous cirripedes and crab. Annals and Magazine of Natural History, (11) **12**: 552-561, pl. 2.

— —, 1951. Some Jurassic and Cretaceous crabs (Prosoponidae). Bulletin of the British Museum (Natural History), (Geology) **1** (6): 171-186, pls. 15-17.

WÖHRMANN, S. & E. KOKEN, 1892. Die Fauna der Raibler Schichten am Schlernplateau. Zeitschrift der Deutschen Geologischen Gesellschaft, Berlin, **44**: 167-223.

WOOD-MASON, J., 1873. On *Nephropsis stewarti* a new genus and species of macrurous crustaceans dredged in deep water off the eastern coast of the Andaman Islands. Annals and Magazine of Natural History, (4) **12**: 59-64.

— —, 1875. On the genus *Deidamia* Willemoes-Smith. Annals and Magazine of Natural History, (4) **15**: 131-135.

— —, 1891. Phylum Appendiculata. Branch Arthropoda. Class Crustacea. In: J. WOOD-MASON & A. ALCOCK, Natural history notes from H. M. Indian marine survey steamer "Investigator", Commander, R. F. Hoskyn, R. N., commanding. No. 21. Note on the results of deep sea dredging during the season 1890-91. Annals and Magazine of Natural History, (6) **8**: 269-286.

— —, 1892. Crustacea. Part 1. In: Illustrations of the Zoology of the Royal Indian Marine Surveying Steamer 'Investigator', pls. 1-5. (Printed by the Superintendent of Government Printing, Calcutta).

WOOD-MASON, J. & A. ALCOCK, 1891. A note on the result of the last season's deep-sea dredging: natural history notes from H. M. Indian Marine Survey Steamer "Investigator": Commander, R. F. Hoskyn, R. N. Commanding, no. 21. Annals and Magazine of Natural History, (6) **7**: 258-272.

— — & — —, 1892. On the results of deep-sea dredging during the season 1890-91. Natural history notes from H. M. Indian Marine Survey Steamer "Investigator", Commander, R. F. Hoskyn, R. N., commanding. Series II, No. 1 Annals and Magazine of Natural History, (6) **9**: 265-275, 358-370, figs. 1-6, pls. 14, 15.

WOODS, H., 1922. Crustacea from the Eocene deposits of Peru. In: T. O. BOSWORTH (ed.), Geology of the Tertiary and Quaternary Periods in the northwest part of Peru: 114-118. (Macmillan and Co., London).

— —, 1925-1931. A monograph of the fossil macrurous Crustacea of England: 1-122. (Palaeontographical Society, London).

WOODS, J. T., 1953. Brachyura from the Cretaceous of central Queensland. Memoirs of the Queensland Museum, **13**: 50-57.

— —, 1957. Macrurous decapods from the Cretaceous of Queensland. Memoirs of the Queensland Museum, **13** (3): 155-175, pls. 4-6.

WOODWARD, H., 1863. On a new macrurous crustacean "*Scapheus ancylochelis*" from the Lias of Lyme Regis. Quarterly Journal of the Geological Society of London, **19**: 318-321.

— —, 1866a. On a new Crustacean, (*Aeger marderi*) from the Lias of Lyme Regis, Dorsetshire. Geological Magazine, **3**: 10-13.

— —, 1866b. Notes on the species of the genus *Eryon* Desmarest from the Lias and Oolite of England and Bavaria. Proceedings of the Geological Society of London, **22**: 494-502.

——, 1866c. Note on a new species of *Ranina* (*Ranina porifera*) from the Tertiary strata of Trinidad. Quarterly Journal of the Geological Society of London, **22**: 591-592.
——, 1867. On a new species of shore-crab, *Goniocypoda Edwardsi*, from the lower Eocene of Hampshire. Geological Magazine, **4**: 529-531.
——, 1868a. Contributions to British fossil Crustacea. Geological Magazine, **5**: 258-261.
——, 1868b. Third report the structure and classification of the fossil Crustacea. Report of the 37th Meeting of the British Association for the Advancement of Science (Dundee 1867): 44-47, pl. 2.
——, 1868c. On a new brachyurous Crustacean (*Prosopon mammillatum*) from the Great Oolite, Stonesfield. Geological Magazine, **5**: 3-5, pl. 1.
——, 1869. Fourth report on the structure and classification of the fossil Crustacea. Report of the 38th Meet. British Association for the Advancement of Science (Norwich): 72-75.
——, 1871. Notes on some new crustaceans from the lower Eocene of Portsmouth. Quarterly Journal of the Geological Society of London, **27** (1): 90-92.
——, 1873. Further notes on Eocene Crustacea from Portsmouth. Quarterly Journal of the Geological Society of London, **29**: 25-31, pls. 1, 2.
——, 1876a. On some new macrurous Crustacea from the Kimmeridge Clay of the Sub-Wealden Boring, Sussex, and from Boulogne-sur-Mer. Quarterly Journal of the Geological Society, **32** (1) (125): 47-50.
——, 1876b. On a new fossil crab from the Tertiary of New Zealand, collected by Dr. Hector, F. R. S., F. G. S., Director of the Geological survey of New Zealand. Quarterly Journal of the Geological Society of London, **32**: 51-56.
——, 1878. On *Meyeria willettii*, a new macrurous crustacean from the Chalk of Sussex. Geological Magazine, **5**: 556-558.
——, 1881. Contribution to the study of fossil Crustacea. Geological Magazine, London, **8**: 529-534.
——, 1888. On a new species of *Aeger* from the lower Lias of Wilmcotte, Warwickshire. Geological Magazine, **5**: 385-387.
——, 1892. Note on a new decapodous crustacean (*Prosopon etheridgei* H. Woodw.) from the Cretaceous of Queensland, Australia. Proceedings of the Linnaean Society of New South Wales, (2) **7**: 301-306.
——, 1896. On some podophthalmatous Crustacea from the Cretaceous Formations of Vancouver and Queen Charlotte islands. Quarterly Journal of the Geological Society of London, **52**: 221-228.
——, 1898. On a new species of brachyurous Crustacea from the Chert beds (Upper Greensand), Baycliffe, near Maiden Bradley Wilts. Geological Magazine, (new series) **5**: 302-303.
——, 1900a. Further notes on podophthalmous Crustacea from the Upper Cretaceous Formation of British Columbia. Geological Magazine, (new series) **7**: 392-401, 433-435, pls. 15-17.
——, 1900b. Note on a Crustacean *Mesodromilites birleyi* gen. et sp. nov. from the Gault of Folkestone, Kent. Geological Magazine, (new series) **7**: 61-64.
——, 1903. On some fossil prawns from the Osborne Beds of the Isle of Wight. Geological Magazine, (new series) **10** (3): 96-99, pl. 5.
——, 1905. Note on a fossil crab and a group of *Balani* discovered in concretions on the beach at Ormara Headland, Mekran Coast. Geological Magazine, **5** (2): 305-310.
——, 1907. On a new brachyurous Crustacean from the "Clypeus-Grit" (Inf. Oolite) of the Cotteswold Hills. Geological Magazine, (new series) **4**: 79-81, pl. 1.

— —, 1911. On a new species of *Eryon* from the Upper Lias, Dumbleton Hill. Geological Magazine, **8**: 307-311.

WOODWARD, H. & J. W. SALTER, 1865. Chart of fossil Crustacea: with descriptive catalogue. (E. Sanford, London).

WRIGHT, C. W., 1997. New information on Cretaceous crabs. Bulletin of the Natural History Museum, London, **53**: 135-138.

WRIGHT, C. W. & J. S. H. COLLINS, 1972. British Cretaceous crabs. Palaeontographical Society Monographs, **126** (533): 1-113.

— — & — —, 1975. *Glaessnerella* (Crustacea Decapoda, Cymonomidae), a replacement name for *Glaessneria* Wright & Collins, 1972 *non* Takeda & Miyake, 1969. Palaeontology, **18** (2): 441.

WRIGHT, C. W. & E. V. WRIGHT, 1945. Notes on some Cretaceous fossils. Geological Magazine, **82**: 126-129, pl. 5.

— — & — —, 1950. Some dromiacean crabs from the English Cretaceous. Proceedings of the Geologists' Association, **61**: 13-27, pl. 1.

WRIGHT, T., 1881 [1882]. On a new astacomorphous crustacean from the middle coral reef of Leckhampton Hill (*Eryma guisei*). Proceedings of the Cotteswold Naturalists' Field Club, **8**: 62.

WÜST, E., 1903. Untersuchungen über die Decapodenkrebse des germanischen Trias. (Jena).

YABE, H. & T. SUGIYAMA, 1935. A new species of the genus *Ranina* (*Lophoranina*) from Haha-jima, Ogasawara Group (Bonin Islands), Japan. Japanese Journal of Geology and Geography, **12** (1-2): 1-4, pl. 1.

YANAKEVICH, A. N., 1977. Middle Miocene reefs of Moldavia: 1-100. (Ministry of Public Education of Moldavia, Soviet Socialist Republic, Kishinev). [In Russian.]

YOKOI, T. & H. KARASAWA, 2000. Additions to Cretaceous decapod crustaceans from Hokkaido, Japan – Part 2. Chilenophoberidae. Bulletin of the Mizunami Fossil Museum, **27**: 191-194.

YOKOYA, Y., 1933. On the distribution of the decapod crustaceans inhabiting the continental shelf around Japan, chiefly based upon the materials collected by S. S. Sôyô-Maru, during the years 1923-1930. Journal of the College of Agriculture of the Imperial University of Tokyo, **12**: 1-226.

YOKOYAMA, M., 1911. Some Tertiary fossils from the Miike coalfield. Journal of the College of Science, Imperial University of Tokyo, **27**: 1-16.

ZIMMERMANN, E., 1944 [imprint 1942]. Krustaceenfunde aus dem samländischen Tertiär im Diluvium von Pommern und Brandenburg. Mitteilungen des Reichsamts für Bodenforschung, Zweigstelle Wien, **62**: 338-343, pl. 20.

ZITTEL, K. A. VON *see* VON ZITTEL, K. A.

ZU MÜNSTER, G. G., 1839. Abbildung und Beschreibung der fossilen langschwänzigen Krebse in den Kalkschiefern von Bayern. Beiträge zur Petrefactenkunde, **2**.

— —, 1840. Über ein neues Brachyurengenus in den tertiären Formationen des nordwestlichen Deutschlands. Beiträge zur Petrefactenkunde, **3**: 23-25.

www.ingramcontent.com/pod-product-compliance
Ingram Content Group UK Ltd.
Pitfield, Milton Keynes, MK11 3LW, UK
UKHW020822290926
15470UKWH00003B/1310

9 789004 178915